Risk: an introduction

Investments, global warming and crossing the road – risk is embedded in our everyday lives but do we really understand what it means, how it is quantified and how decisions are made? Explaining the concepts, methods and procedures for risk analysis, Bernardus Ale casts light on the often overlooked basics of this fascinating field. Theoretical concepts are applied to everyday examples, providing a clear and coherent exploration of risk perception, assessment and management.

Developed from the Safety Science or Risk Science course taught at Delft University, *Risk: an introduction* serves as an excellent starting point for students at undergraduate and postgraduate level, as well as those wishing to develop a career in risk management and decision-making.

Bernardus Ale is Director of Research at the Netherlands Institute for Fire and Disaster Abatement (NIBRA). He is also Professor in Safety and Disaster Abatement in the Faculty of Technology, Policy and Management at Delft University.

Risk: an introduction

The concepts of risk, danger and chance

Ben Ale

LONDON AND NEW YORK

First edition published 2009
by Routledge
2 Park Square, Milton Park, Abingdon, Oxon OX14 4RN

Simultaneously published in the USA and Canada
by Routledge
711 Third Avenue, New York, NY 10017, USA

Routledge is an imprint of the Taylor & Francis Group, an informa business

Typeset in Sabon by
GreenGate Publishing Services

British Library Cataloguing in Publication Data
A catalogue record for this book is available from the British Library

Library of Congress Cataloging in Publication Data
Ale, Ben J. M.
Risk : an introduction the concepts of risk, danger, and chance / Ben Ale. --
1st ed.
p. cm.
Includes bibliographical references and index.
1. Risk management. 2. Decision making. I. Title.
HD61.A418 2009
658.15'5--dc22 2008044873

ISBN10: 0-415-49089-8 (hbk)
ISBN10: 0-415-49090-1 (pbk)
ISBN10: 0-203-87912-0 (ebk)

ISBN13: 978-0-415-49089-4 (hbk)
ISBN13: 978-0-415-49090-0 (pbk)
ISBN13: 978-0-203-87912-2 (ebk)

Contents

LIST OF TABLES AND FIGURES

Tables

Figures

PREFACE

For years the course on Safety Science or Risk Science at Delft University has been given from overheads and Microsoft PowerPoint files. No written course notes were developed after the early 1980s. The people for whom the courses were and are given are usually complete novices both in the concepts of risk and decision making and in the associated mathematics and other methodologies. This leads to the question of where to begin.

The concept of risk is not very new. Its earliest mention is over two millennia old, and the concepts of probability and consequence were developed in the seventeenth century. However, probability in particular proves difficult to grasp. And as safety and risk are close to our heart, it is difficult to separate science and technology from emotion. In fact, this difficulty has developed into a field of science in its own right.

This book provides an introduction to the field. Some concepts are worked out in more detail, either because relevant material is difficult to find elsewhere or because it illustrates the point. On other occasions – in particular, for the more fundamental underlying mathematics – the reader is referred to other literature.

Many people have helped develop the material and ideas in this book. Special thanks are due to Linda Bellamy, who over the past 20 years has sorted out many data problems that supported the further development of risk analysis methodology, as well as my English; Roger Cooke and Dorotha Kurowicka, who converted Bayesian belief nets into a usable analysis tool; Andrew Hale, whose chair I was allowed to take, which led to the need for course material and this book; and Ioannis Papazoglou, who sticks to justifiable mathematics, and rightly so.

Finally, very special thanks are due to Marga van der Toorn, my wife, who saw many holiday days become working days, but kept her cool.

1 Introduction: what is risk?

Risk is everywhere and always has been. Although industrial risks, environmental risks and health risks seem new, they have been around since the origin of humankind. Human beings have always tried to be safe; that is, to maximize their safety, or at least the feeling of safety. Therefore, the aim has always been to minimize risks and manage them where possible. Safety and risk are like conjoined twins; they cannot be separated. Safety is related to the absence of risk, but risk does not have to be absent for a situation to be called safe. The risks of modern technological society can be managed by using the means society has developed.

But today the decision to reduce risk is political in principle, just as it always has been. This means that decisions about activities, and especially risky ones, always involve what we think about society, about the value of human life, about democracy, about ourselves – in short, values. In many instances, these other values and opinions are in the end much more important for the decision than the risk. This, however, does not take away the need to get the science right.

Therefore, this book is primarily about risk. The first step in understanding decisions about risk is to understand risk itself. In the final chapter of this book we shall look at decision making and at what is involved in what is called risk governance.

Many of the examples in this book involve risks to human life and health. Decisions on risks involving life and death are much closer to our heart than decisions about money, and examples from this area are more enlightening about the problem that a risk analyst, assessor, manager or governor is facing, but we shall see that the concept of risk is everywhere.

INTRODUCTION

It is said that modern society is a risk society (Beck, 1986). And indeed some risks are new. Moreover, because of the global connectivity of our societies, many risks are shared by all of us. Even so, many risks in former times were just as dangerous to the society they threatened. They constituted a threat to the whole of the known world, and all known societies were exposed to the danger. For example, between 1347 and 1350 the Black Death wiped out one-third of the population of Europe (www.20eeuwennederland.nl). In the seventeenth century the average life expectancy was 25 years, and to reach the age of 45 was exceptional. As recently as 1918 the Spanish flu killed 170,000 people in the Netherlands alone.

What now is called industrial risk also has roots in the past. Pliny the Younger described illnesses among slaves (Ramazzini, 1700). In 1472, Dr U. Ellenbog of Augsburg wrote an eight-page note on the hazards of silver, mercury and lead vapours (Rozen, 1976). Ailments of the lungs found in miners were described extensively by Georg Bauer, writing under the name Agricola (1556). In the seventeenth century a significant proportion of the crews of ships sailing to the East and West Indies never made it home.

The Netherlands has a long history of having to deal with the threat of floods. In the Middle Ages, several groups, such as Huguenots and Jews, fled to the Netherlands because of oppression by the government of their home country. These people descended from the Central European Plain into the Low Countries, the swamp that is now the Netherlands. For a long time, the only authorities that were accepted were the 'water boards'. These were deemed necessary to manage the flood defences. The oldest water boards were those of Schieland (1273), Rijnland (1286) and Delfland (1319). Today the Netherlands has 478 people per square kilometre, making it one of the most densely populated areas in the world, and houses the harbour of Rotterdam, Schiphol Airport and a third of the refinery capacity of Europe, and hence managing the risks resulting from the close proximity of people and industry has become just as important an activity as managing the risks of flooding.

Attempts to avoid unnecessary risk have also been part of human activities for as long as history has been written. Those who had something to lose surrounded themselves and their possessions with walls, castles, guards and armies. Those who had enough money left the city to escape the plague (Chaucer, writing in the fourteenth century). And societies have long put people into power, for example a king, in order to protect them from harm.

Nevertheless, it is true that worldwide and in absolute numbers the number of disasters and the associated costs have increased. At the same time, the population of the earth is increasing, suggesting that people are increasingly living in less and less suitable locations (OECD, 2003).

This raises the question of why risk management looks so different today and why we have so much difficulty in formulating an organized policy on risk, whether we are in public office, in government or in private enterprise. To understand the reason, we have to understand risk and why we sometimes gladly take risks and at other times seek to avoid risk. We have to understand how to measure and assess risk. We need to have a knowledge of risk assessment methodologies in order to understand the genesis of accidents and to be able to devise strategies to eliminate them or reduce the probability of their occurring. We have to understand the forces involved in the debate about risks and we have to understand the mechanics of making decisions concerning risk. Therefore, we need to be aware of the development of risk perception research and findings, and we have to have insight into the many value-laden choices that people make. Which means that a study of risk is, in many ways, a study about ourselves.

The risks discussed so far are concerned directly with life, health or death. There are many other kinds of risks, such as losing money in a casino or on the stock exchange, or the destruction of crops by bad weather. We shall see that the techniques of analysing, characterizing and managing these risks are common to all these types of risks. But the risks that are most widely discussed and debated are those associated with a threat to our own health. And even if it seems to be 'only money', we do become really interested if there is a threat to the money we use to buy food and shelter. Therefore, most of the examples in the following will have to do with life and death. But lesser risks will be referred to as well.

WHY WE TAKE RISKS

With the possible exception of those who are 'thrill seekers', we rarely take risks for the sake of it. We accept risks as part of an activity we undertake – an activity we undertake because we have to or want to.

Some activities are necessary. We have to eat. Therefore, we have to acquire food and drink. In prehistoric times we would go hunting in forests and on plains inhabited by creatures that would try to eat us before we could eat them.

Only for some 5,000 years or so have we obtained our food by farming. But farming is itself not without dangers. Today we use machines, build houses, mingle with traffic as drivers or pedestrians, use refineries, build chemical plants and use nuclear power.

Today, most of our activities do not seem to have a direct relationship with survival. But still, no money, no food. So, the act of earning money, whether in agriculture or in a chemical factory, ultimately is earning the means of living.

Human needs and wants can be divided into several layers or classes. Maslow (1943) conceived these as taking the form of a mountain or pyramid (Figure 1.1). The base of the pyramid represents our needs for survival and

Figure 1.1 Maslow's hierarchy of human needs

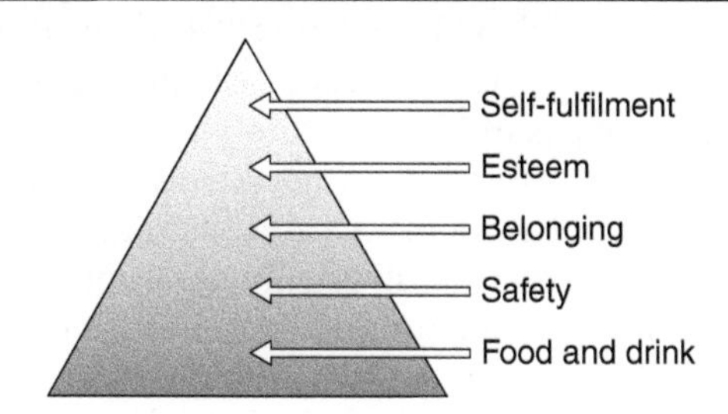

Source: Adapted from Maslow (1954)

the top represents what Maslow thought were the higher needs of humankind: esteem by others and self-fulfilment. To fulfil our needs we use technology. Unfortunately, we humans are far from perfect, and so is our technology. Failures of ourselves and of our technology infringe on the needs of other people to fulfil their needs and wants. This is usually not what we want ourselves. Therefore, these failures are called unwanted side effects or unwanted effects. Since these unwanted effects may bring us into conflict with other people, we want to prevent them. And if we cannot prevent them, we at least try to minimize them and perhaps recognize the need to compensate other people for any damage or nuisance we have inflicted on them.

Interestingly, in the overwhelming majority of cases, failures of ourselves or our technology harm only ourselves. Usually there is not even any damage to our health or any danger to our lives. Most failures merely mean that the technology or the system we are using fails to work. We have already seen some examples of how we can investigate what the consequences of failure could be, but sometimes it is extremely simple: if a car does fail (that is it, as long as it is not hit by another car) it just sits there and we complain. If an aircraft stops working, it falls to the ground. Gravity takes care of that. A crashing plane, by contrast, is usually bad news for anybody who is in it; it means something much worse than just not being able to get home on time. Reliability has to do with both cases; safety only with the latter. Clearly, whether a failure is a safety issue is a matter of what sort of failure in what system.

THE ESSENCE OF RISK

The essence of risk was formulated by A. Arnaud as early as 1662 (Figure 1.2): 'Fear of harm ought to be proportional not merely to the gravity of the harm, but also to the probability of the event.'

The essence of risk lies in the aspect of probability, chance, uncertainty. Even the most hazardous system or the most dangerous activity is not certain to bring us harm or death. Only one activity will definitely get us killed, and this is an activity that we would normally consider quite safe: living itself. Note that Arnaud treats probability more as a certainty than as an uncertainty and as, in principle, measurable. Many who have followed him have done the same. Allan H. Willett (1901) defined risk as 'The objectified uncertainty regarding the occurrence of an undesired event' and Frank Knight (1921) as 'measurable uncertainty'.

There are many risks that do not involve human life or health or the state of the environment. In many cases the stakes are of a different nature. In the financial markets, risk is usually associated with losing money as a consequence of investments turning bad, mortgages not being paid back or fraudulent bookkeeping. In construction the risks are associated with, say, completing a railway on time and within budget.

Figure 1.2 Page 467 of Arnaud's book *La Logique ou l'art de penser*

IV. PARTIE. 467

trente mille fois plus probable pour chaque particulier qu'il ne l'obtiendra pas, que non pas qu'il l'obtiendra.

Le defaut de ces raiſonnemens eſt, que pour iuger de ce que l'on doit faire pour obtenir vn bien, ou pour éviter vn mal, il ne faut pas ſeulement conſiderer le bien & le mal en ſoy, mais auſſi la probabilité qu'il arrive ou n'arrive pas; & regarder geometriquement la proportion que toutes ces choſes ont enſemble: ce qui peut eſtre eſclairci par cét exemple.

Il y a des jeux où dix perſonnes mettant chacun vn eſcu, il n'y en a qu'vn qui gaigne le tout, & tous les autres perdent: ainſi chacun n'eſt au hazard que de perdre vn eſcu, & en peut gaigner neuf. Si l'on ne conſideroit que le gain & la perte en ſoy, il ſembleroit que tous y ont de l'avantage: mais il faut de plus conſiderer que ſi chacun peut gaigner neuf eſcus, & n'eſt au hazard que d'en perdre vn, il eſt auſſi neuf fois plus probable à l'égard de chacun qu'il perdra ſon eſcu, & ne gaignera pas les neuf. Ainſi chacun a pour ſoy neuf eſcus à eſperer, vn eſcu à perdre, neuf degrez de probabilité de perdre vn eſcu, & vn ſeul de gaigner les neuf eſcus; Ce qui met la choſe dans vne parfaite égalité.

V vj

Source: Gallica (www.gallica.brf.fr)

All these examples have in common that the outcome of an action, a decision or an activity is uncertain – sometimes more uncertain than at other times. In order to deal with uncertainty in an organized way, we use the concept of probability. Probability is the chance that something will happen. Probability can be calculated, estimated, judged, believed in. Uncertainty and our judgement of it play an all-important role in the way we deal with hazards and risk. Organized, institutional dealing with risk is also referred to as risk management.

Risk therefore is a combination of consequences and probabilities. In Arnaud's view the true measure of risk is the product, in a mathematical sense, of probability and consequence. Risk is chance multiplied by effect – what in mathematical terms is designated the expectation of the consequences. We shall see that decisions that follow Arnaud's rule in having the acceptability of an activity directly proportional to this measure of risk are common in economics. However, the more contentious decisions – and these are often related to issues of life and death – do not seem to follow this rule. Many attempts have been made to capture apparently different relationships between acceptability, probability and consequence. We shall see some later.

The magnitude of risk

Arnaud mainly performed risk analysis, comparison and decision making in relation to the Casino of Paris, so he was not too concerned with societal and political discussions. In particular, discussions about the use of nuclear power, about the risks of chemical industry and the associated transport of hazardous chemicals, and about the long-term effects of human activities on climate have shown that decision making is often to do with more than consequence and probability alone (Gezondheidsraad, 1993). As we shall see, there are many factors besides these that influence the decision. Therefore, the process of risk management can be summarized as in Figure 1.3 (van Leeuwen and Hermens, 1995).

After identification of all the potential adverse events, the probabilities and consequences are modelled and quantified. The risks are also qualified. Qualification in this context means establishing other attributes of the activity with which the risk is associated and which are important for the decision to undertake the activity. These attributes are often value laden, especially when the risk involves potential harm to human life or health. Although it may seem that establishing the magnitude of risk is value free, it often is not, because, as we shall see later, the way this magnitude is expressed may itself contribute to the framing of the decision. After this work has been done, the information is ready for use in a decision-making process. After it has been decided whether the risk is acceptable or has to be reduced, the risk is monitored and a new cycle may start, depending on whether the risk still seems acceptable or not.

Figure 1.3 Steps in the risk management cycle

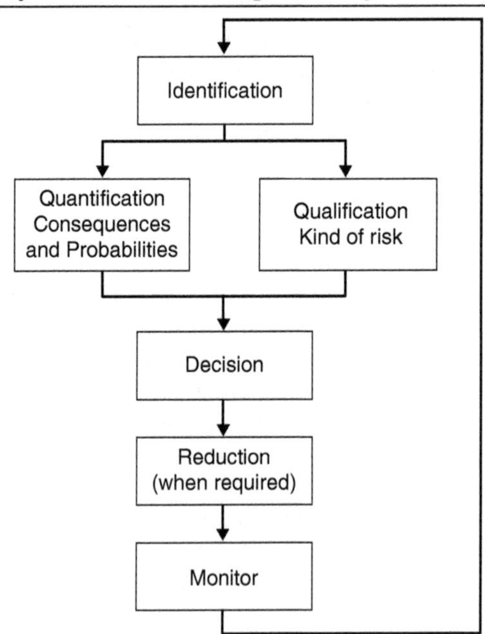

Although decision is only one of the blocks in Figure 1.3, it is making the decision that usually takes the most time and requires the most discussion.

In real life the risk management process is not as clear-cut as the schematic diagram suggests. As has been said, value judgements are often made in the steps where information is assembled, and in this way the gathering and presentation of information becomes a part of the decision making. In real life, however, the people who gather and assemble the information are more often than not different from the people who make decisions. The information is often gathered by consultants and advisers. The decision is often taken by policy makers and politicians. At least, in a democracy we expect elected politicians to make these decisions. Therefore, everybody involved in risk management needs to be conscious of his or her own value judgements and how they influence the outcome of the process, and must try to keep the real decision making where it belongs. As Harry Otway (1973, 1975) put it:

> Risk estimation may be thought of as the identification of consequences of a decision and the subsequent estimation of the magnitude of associated risks. Risk evaluation is the complex process of anticipating the social response to risks … this could be termed as the 'acceptability of risks'.

We could also make a distinction between risk management and risk governance. Risk management may be thought of as keeping risk within defined limits against defined costs. Risk governance includes the process by which we deal with a problem that involves risk, but also many other things.

Risk and expectation

As we have seen, in its simplest form the magnitude of risk is the total value of the expected outcomes, probability multiplied by consequence:

$$R = p \times c$$

So, if the prize in a lottery is €1 million and the probability of winning it is 1 in a million, then the value of the risk is

$$R = p \times c = 1{,}000{,}000 \times 0.000001 = 1$$

Such an expectation can be used to judge whether it is worthwhile to buy a lottery ticket. If the ticket is more expensive than the value of the prize, we could decide not to buy the ticket. Later we shall see that in practice it is not as simple as that. But in this case the cost–benefit relationship is straightforward.

In a real lottery, however, there is not one prize, but a range of prizes. If the probability of winning any one of these prizes were the same, the value of the prizes could be calculated in the same way as the expectation value of the throws of a dice. This is calculated by summing the values of the separate prizes or outcomes, that is, the magnitude of the outcome multiplied by the probability.

A dice has six faces, numbered 1 to 6, and the probability of throwing any one of these numbers is 1/6. Thus, the value of the 'risk' is

$$R = \sum_{i=1}^{6} P_i \times c_i = \sum_{i=1}^{6} {}^{1}\!/_{6}\, i = (1 + 2 + 3 + 4 + 5 + 6)/6 = 3.5$$

If the probabilities for the different outcomes are different, then the risk in general is

$$R = \sum_{i=1}^{6} P_i \times c_i$$

This definition of the value of risk is used in finance and insurance. Using this definition of the risk magnitude in decision making means assuming that probability and consequence are the only two dimensions that are important in the decision. As was mentioned earlier, in many instances this is not the case.

Immeasurable risk

Now suppose we are confronted with a risk that cannot be measured, for instance the risks associated with global warming. Global warming presents a tricky problem for the management of risk. We cannot really assemble data on the effects before they take place, and we cannot assemble data on the probability either. The reason is very simple: there is only one earth and we cannot go back in time. So, probability in the frequency sense is not an issue here. When and if certain effects take place, there will be absolute certainty: the effects will have taken place. And before that there is only probability in the true Bayesian sense: a degree of belief that certain negative – or positive – effects will result from our activities. The degree of belief may be increased by factual information, but it remains a belief.

In such cases, where the 'proof' by mathematics or by experiment is impossible in principle, decision making is much more difficult, and having a decision implemented if the cooperation of non-believers is involved is more difficult still. In the absence of actual facts, values and interests dominate the decision-making process. When involved in a decision-making problem, a wise scientist always first tries to assess whether the problem is about risk or about uncertainty and fundamental absence of knowledge. The former problem can be dealt with by science, the latter only by discussion and conviction. Or crusade.

SAFETY SCIENCE

Safety science deals with risk problems. Risk problems can be assessed and explained. Risk can be studied, analysed, prevented and managed. There are technologies and methodologies that can be used and applied.

Risk problems are never isolated. They have a context in society and in law. There is policy and planning.

Risk science is all about choices, and the methods and tools to make these choices.

After their introductory course the student who uses this book as his or her course material will:

- know how to explain what sort of risk problem is at hand, and how serious it is;
- know what the principles of studying, analysing, preventing and managing risk problems are;
- be able to describe and use available safety methods, methodologies and techniques;
- be able to describe the legal and societal framework in which safety policy and planning occur;
- understand the role of beliefs, values and perceptions in decision-making processes.

Risk is, in a way, the opposite of safety, but not its exact opposite, as we shall see. But the conjoined twin of risk is so important that we have to discuss safety a little more first.

What is Safety?

Everybody knows how it feels to feel safe – and how uncomfortable it feels to be in danger, to feel threatened. Yet nobody really knows what safety is. Safety is something like not being in danger, just as health is the absence of illness. Safety, health and the environment are generally considered as things worth protecting and cherishing. They are 'good' things. As I have said, safety is very dear to our heart. It is considered one of our basic needs, just after food and drink (Maslow, 1943, 1954). We never can have enough of it. And that is where the problem starts. The quest for safety can come into conflict with one of our other basic needs, or with the needs of others. And suddenly we need a way to decide, to choose. And thus we need a way to measure. And we cannot, because we do not know what safety really is. Therefore, we cannot answer the question 'how safe is safe enough?' other than by saying 'when we feel safe'. Such an answer leads to obvious problems if decisions are to be taken by people other than ourselves or if our need to feel safe can only be satisfied by others. When have they done enough? And what if doing 'enough' infringes on their needs? And what if you feel safe but your neighbour does not, and making the neighbour feel safe costs YOU money. Again, what if your neighbour wants to do something that makes YOU feel unsafe? The neighbour gets the money and you the danger. Should the neighbour stop? Or should he or she give you money? Because feeling safe is so important, and there are so many things that make us feel unsafe, safety, like health and the environment, is the subject of public debate, societal processes and political decision making. We cannot be safety scientists without looking at these processes.

If we Google 'safety', the free dictionary comes up with the following definition:

> SAFETY is the condition of being safe, freedom from danger, risk, or injury.

The first part of this definition is not very helpful: safety is to be safe; we know that. The second part is more interesting: freedom from danger, risk or injury. These three, danger, risk or injury, can be defined in 'operational terms'. This is important, because using operational terms means that we can observe, measure and thus weigh. And as soon as we can measure and weigh, we can choose and decide.

Danger and injury are different from risk – and for decision making the difference is important.

Danger

Danger is also called hazard or threat. Dangers are the things that make us feel unsafe. The world is full of dangers. These dangers are associated with activities: our own and those of others. Examples of these activities are

- driving a car
- transporting fuel and chemicals
- having a supertanker full of oil cross the ocean
- producing chemicals
- storing chemicals
- having a nuclear power plant
- flying
  - being in an aircraft
  - having aircraft flying over our head
- pollution of the sea
- burning fuel (and producing gases that result in global warming)
- storing fireworks
- DIY
- going up and down stairs or ladders.

Such a list makes one wonder how anyone can ever feel safe. In order for us to feel safe, we have to keep these hazards under control – not only the big, spectacular hazards on a national or world scale, such as tsunamis and chemical disasters, but also the small, everyday hazards.

Dangers or hazards have two properties or dimensions that need to be considered:

- What is the nature of the potential damage?
- How bad is it?

As an example, consider standing on a ladder (Ale, 2006). The hazard is falling from the ladder to the ground. Depending on the height, you may be injured or killed if you fall. So, the nature of the damage is mechanical damage to your body, and that damage can be bad enough to cause death.

As another example, consider having a chemical plant using methyl isocyanate. This chemical is a vapour that causes chemical burns when in contact with the mucous membranes of our body. Mucous membranes are the moist surfaces of our body such as the cornea and the inner lining of our lungs. Attacking these leads to blindness and to oedema of the lungs. The latter can be lethal. In 1984 in Bhopal in India a few tons of this chemical was released from a pesticide plant to the atmosphere. Three thousand people were killed and 200,000 injured, many of whom died later (Bahwar and Trehan, 1984). This example clearly demonstrates the hazards associated with the use of certain chemicals.

The description of the hazard associated with the use of a ladder already hints at the problem with using hazard as the basis for making a decision. What should be the determining consequence in our decision? Breaking some bone, or death? Or the fact that we mostly survive the use of a ladder in one – undamaged – piece?

Injury

An injury is physical damage to the body, such as broken bones, wounds, or burns from chemicals or heat. Injury is different from illness. Illness is a malfunction of the body not caused by physical damage. Illness is caused by bacteria, viruses, bad food, insufficient or excessive food, an unhealthy lifestyle, etc. Preventing and curing illness is often considered part of healthcare rather than of safety science. However, the probability of contracting these illnesses is a risk, just as for any of the risks mentioned earlier, and therefore health risks can be managed via similar processes.

Risk

Risk is the entity that is mostly used in managing hazards, dangers and injury. Keeping hazards and injury under control is called risk management. Systems designed to manage risk are often called safety management systems (SMS).

Risk has two dimensions: extent of consequence and probability of occurrence. This gives us the possibility of solving the problem that we encountered with the ladder example. Rather than considering only one possible consequence, we consider the whole range of possible outcomes of our activities – the use of a ladder is an example – together with their probability. We can then see that not having an injury at all is the most probable outcome. Even when we do have a fall, the probability of survival is still very high, as can be seen from Table 1.1 (Papazoglou *et al.*, 2006). The decision to use a ladder may, in this light, be driven more by the probability of survival than by the seriousness of the most adverse consequence: death. Similarly, the large policy decisions

Table 1.1 Risk of using a ladder

Consequences	Probablility
None	Once in 1 year
Recoverable injury	Once in 1,724,137 years
Permanent injury	Once in 3,571,428 years
Death	Once in 55,555,555 years

such as on the siting of chemical industries or the growth of air traffic are made considering the whole range of consequences and probabilities.

The concept of risk is not confined to questions of life and death. It is similarly useful in the area of financial risk management (McCarthy and Flynn, 2004). In fact, it was in economics – and in the casino – that many of our ideas about risk and risk management originated. But debates about decisions involving life and death are often much more heated, and the issues are considered much more political, whereas decisions involving financial risks are considered to be far more of a technical nature. As we shall see later, decisions are never really based on consequence alone. There is always some element of probability in the decision.

SECURITY AND SAFETY

Although 'security' as a word has a similar meaning to 'safety' and there is hardly any difference between feeling secure and feeling safe, in the practice of risk management these two terms have different meanings. Safety is used when the threat is an unwanted side effect of something else we want. Safety thus is associated with incidents and accidents. Security deals with malicious acts, such as sabotage and terrorism.

There is a grey area, however, where the distinction between security and safety, between accident and criminal act, is difficult to draw. This is where accidents are the result of management decisions or decisions of government. The decision not to stop the storage of fireworks in the city of Enschede ultimately cost 22 lives (Ale, 2001) and caused half a billion euros worth of damage. Is that an accident, an unwanted side effect of a desired activity, or was it the result of criminal negligence?

In these areas, safety, security, politics and law form a complicated union that the safety scientist needs to understand in order to carry out his or her job responsibly. This understanding also helps to give an insight into the limitations of safety science and the safety scientist in influencing decisions. It helps therefore to take a position based in science and in ethics. As I have already pointed out, pursuing safety can have a negative impact on other things people pursue, such as health and wealth. The playing field of risk decisions is seldom level, and the safety scientist has to find a position in this field.

TECHNOLOGICAL RISK

Technological risk is present wherever systems are operating in which technology and people interact. Sometimes the risk is to people operating the machines and equipment, where they are using technology to achieve a particular goal. Sometimes it is because the activity is located near to so-called third parties. Third parties are people who do not take part in the activity but are exposed to the risks. These could, for example, be people living around a facility, or near a route used for the transportation of chemicals or fuels; people living under the flight paths of an airport; or people who work in the office of a chemical factory. It could also be passengers using a particular means of transport, or pedestrians.

Technological risks are only one kind of the risk humankind has to face. Figure 1.4 classifies many common risks according to the extent of the damage they involve and the probability of their occurrence.

The major technological risks associated with explosions, toxic clouds, etc. are located in the right-hand bottom corner of the figure. As can be seen, there are risks for which the probability of an event is much larger, such as traffic accidents and accidents in the workplace such as falls. Famines and epidemics are still the causes responsible for the majority of untimely deaths, followed by floods, earthquakes and war (Nash, 1976). Famines and epidemics occur much more often than technological disasters. These sorts of events tend to have larger consequences, with numbers of victims in the order of tens of thousands. Wars and conflicts sometimes result in millions of victims. The events that the safety scientist deals with are those which have a technological edge. It is the use or misuse of technology that creates a situation in which lives can be at stake. Safety scientists should be aware that their problems may be judged minor with respect to other risks. On the other hand, their problems may have widespread political and societal consequences. These varying opinions will be investigated in chapter 6.

Figure 1.4 Risks divided according to their consequences and probabilities

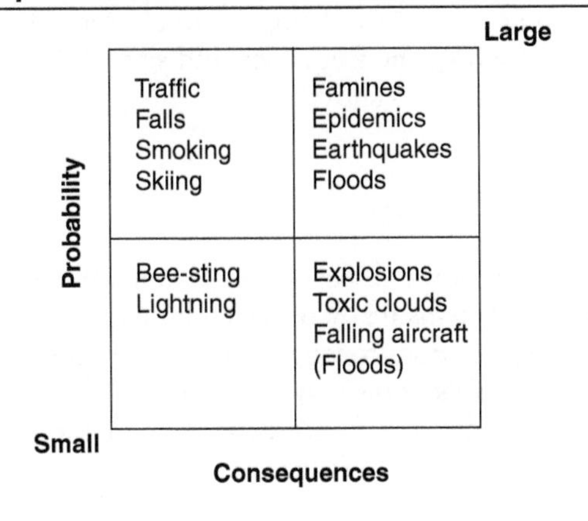

Figure 1.5 Number of natural and health disasters reported in the world

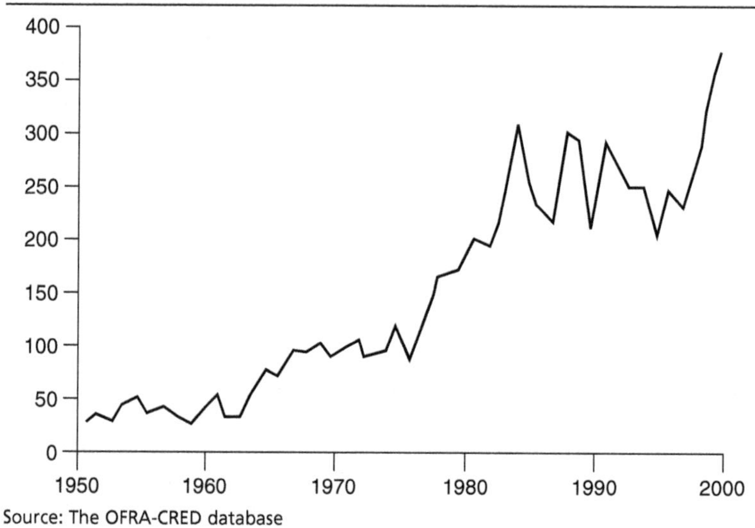

Source: The OFRA-CRED database

In any case, there is a continuous stream of man-made accidents with many victims, only some of which are caused by intentional malicious acts. Most accidents genuinely are accidental.

There has been a continual stream of accidental explosions: Delft (1654; 1,500 casualties), Leiden (1807; 51 casualties (NN, 1933)), Oppau (1921, 600 casualties (Nash, 1976)), Enschede (2000, 21 casualties (Ale, 2001)). Aircraft accidentally crash into inhabited areas such as Amsterdam (1993 (www.aviation-safety.net)), Turin (1996), Kinshasa (1996) and the crash in New York in November 2001, which because of the events of 9/11 almost became obscure.

Also, trains crash into each other and into buildings. Accidents such as those in Harmelen (1962) (Jongerius, 1993), Ladbroke Grove in London (1999) (HSE, 2000) and Tokyo (2005 (www.cnn.com)) subsequently led to new policies, and to new measures being taken to enhance safety.

In the world as a whole the frequency and size of disasters are increasing (OECD, 2003), as Figure 5 shows.

RISK SCIENCE DEFINED

Risk science is the study of the threats posed to humankind and the environment by technological systems many of which human beings are a part. Risk science tries to find explanations for the occurrence of unwanted events. From these it tries to make models of these systems with the focus on describing the risk, which helps to predict the future behaviour of the systems and helps those involved to find measures to reduce the risks posed by these systems.

Risk science also deals with how judgements about the acceptability of risks, on criteria, norms and standards, come about and how policy decisions are made and justified. Risk science aims at making knowledge and insights operational in managing risks and limiting avoidable damage, with the aim of improving life and making people safe (and making them feel safe).

The relationship between safety science and risk science is even closer than that of conjoined twins. They are one and the same.

2 Modelling risk

In this chapter we concentrate on establishing the magnitude of risk. To do so, we will at some point need a risk metric or metrics. The choice of metrics is not necessarily value free, as we have already seen. Some people believe that risk cannot be measured at all. In this chapter we shall take risk as being measurable, at least in principle, however.

As we have seen in the previous chapter, risk has two dimensions, extent of the consequences and probability of the risk's occurrence. This makes it difficult to observe or measure risk directly. If there is a chance that something bad can happen, there is a certainty that it will happen, sooner or later. This certainty is also called 'Murphy's law':

> If anything can go wrong, it will.

We just do not know when. And if the probability is small, maybe we will not actually see it happen in our lifetime. In such a case the uncertainty about when is so large that from the point of view of a human, the uncertainty about when it will happen is virtually the same as uncertainty about whether it will happen at all. In the case of such large uncertainties and the impossibility of gaining sufficient information from direct observation, we need models. We need models even more when we want to predict in advance the result of an intervention we are considering making in the flow of events.

Murphy's law

Murphy's law originated at Edwards Air Force Base in 1949. Captain Edward A Murphy was heading a team. When one of his technicians made a mistake, Murphy was reported to have said, 'If there is any way to do it wrong, he'll find it.' This law was later rephrased to the formulation that currently is known as Murphy's law: 'If anything can go wrong, it will.'

CAUSE AND CONSEQUENCE

If we need to do something about the possibility that certain events will take place, we have to assume that there is something we can do. We have therefore to assume that an unwanted event – once it has happened – has a cause and that we can intervene between cause and effect (Figure 2.1).

Figure 2.1 When we want to intervene, a consequence has to have a cause

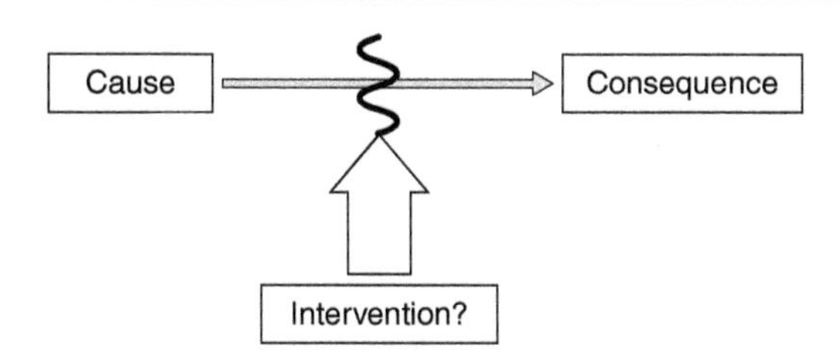

This is an important assumption, because causes are also events, and therefore this assumption implies that we can follow back in time the trace of consequences and causes endlessly. This notion is even the basis of Einstein's theory of relativity: a cause always precedes a consequence, no matter how you look at it (Einstein). This excludes sudden, spontaneous events. Even if an event seems to have no cause – sometimes referred to as 'an act of God' – it has to have one. All we have to do is to find it.

But what about events that have not happened yet; say a potential future accident? Such an event may well have more than one potential cause, and which of them is going to be the actual cause we do not know. We may be able to establish the probability that one factor or the other is going to be the cause. Similarly, after the event, when we have a list of possible causes, but cannot (yet) decide definitely between them, we may be able to assess the probability that one or the other has been the actual cause.

In many cases the cause consists of multiple events. Two or more things have to go a certain way before the next event can happen. As an example, take a car light (Figure 2.2).

Every light in a car is connected to the battery AND the generator by means of a switch. Several events can cause the light not to be on, or to fail to work. The light bulb may be broken, the switch may be turned off, or BOTH the battery AND the generator may fail to supply energy. These latter events both have to happen for the light not to work. We will see later how we can use formal analysis techniques to build a logical model for this and other systems and how we can deal with probabilities in such systems.

Figure 2.2 A car light system

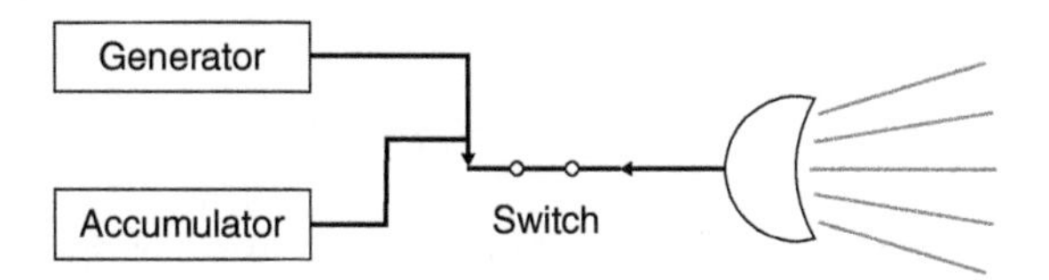

Reality is always much more complicated than we can model. In reality, everything is related to everything else. When looking at reality in this way, one can rarely if ever speak of an identifiable cause. Nevertheless, if certain combinations of events usually are followed by certain combinations of other events, we consider that we have found the cause (Kant, 1783). We have to if we want to be able to intervene sensibly, rather than accept the world and our fate as they are.

Whereas according to our way of reasoning, events always have to have a cause, events do not always have to have consequences. They might, but they do not have to. If an apple comes loose from the branch of a tree, it falls towards the ground. It does not have to reach the ground. Somebody (maybe Newton) may catch it before it actually hits the ground. In such cases there is a probability smaller than 1 that the event – in this case apple hits ground – will happen as a consequence of the cause. A probability of 1 means certainty. We will have a closer look at probabilities later.

An event may also have more than one consequence. Some of these may be mutually exclusive. If the brakes of a stationary car on a slope fail and the car moves, it either goes forwards or it goes backwards. It cannot do both simultaneously (unless the car falls apart). Other consequences may occur together, and are thus not mutually exclusive. If you fall from a roof, you may break your arm AND your leg.

We should be aware, though, that some processes are so complicated and the chain of events so subject to chance that we will not be able to find the cause, no matter how hard we try. A thunderstorm in Europe may ultimately be caused by the motion of the wing of a butterfly in China, but how could we ever find that out? Such phenomena are also called chaotic (Stewart, 2002). That is not to say that there IS no cause, but for all practical purposes we may treat the events as if they do not have one. For these events it is not possible to intervene between cause and event. We have to accept the events as they are. We may, however, still be in a position to do something about the consequences.

PROBABILITY

Just as there is a discussion about what cause really means, so there is a discussion about probability. Roughly speaking, there are two schools of thought: the frequentists and the subjectivists. The frequentists take probability to mean the number of specified occurrences divided by the number of tries:

$$P(A) = \frac{N_A}{N}$$

in which P(A) is the probability of event *A*, and N_A is the number of occurrences of *A* in a total of *N* events – in the case of dice, the number of times a particular number (say 6) is thrown divided by the number of throws. The outcome will be 1/6 if the number of throws is big enough. Frequentists do not care whether in an actual experiment, where the number of throws necessarily has to be small, the number of 6's is not one-sixth of the total number of throws.

Subjectivists, however, say that the frequentists only *believe* that the probability of throwing 6 is 1/6, and that there is no way of knowing for certain. First of all, nobody has done an experiment with an infinite number of throws, which is the hard mathematical demand for the ratio to really be 1/6. And, they say, how do you know that the next throw will behave the same way as all previous throws? You cannot possibly know the future. So, the subjectivists say that probability is a measure of belief that a future event will turn out in a certain way. One of the founding fathers of the subjectivists was Thomas Bayes. He wrote an essay, published after his death, that addressed the – largely philosophical – problem of assessing the outcome of a future event given knowledge of past events (Bayes, 1763). Therefore, subjectivists are also called Bayesians.

Bayes developed a method to combine beliefs, which may be pre-existing, with observations or new information as follows:

Say we have an event A for which we want to know the probability. We have a pre-existing belief that this probability equals $P(A)$. We now make observations or find evidence or develop a new belief that leads to a probability $P(E)$ (E for evidence). Now the new probability of A is the probability of A given the evidence E, which is noted as $P(A|E)$. We now need to know one more thing: the probability of E given A. In other words, if A is true or occurs, what then is the probability that the evidence occurs or is true also, or $P(E|A)$? This does not need to be true. Think for instance of tests for a certain illness. The test may be positive for people who are not ill and may be negative for people who are ill.

Say that the event is that the patient is ill and E is that the test is positive. Say the probability of a wrong test when the patient is ill is 1 per cent, and that the probability of a wrong test when the patient is not ill is 2 per cent. This means that 99 per cent of the patients who are ill will be detected and 2 per cent of the healthy people will test as ill.

$$P(B|A) = 0.99$$
$$P(B|A^*) = 0.02$$

where the * means the opposite of A. In this case, therefore A^* means that the patient is not ill and E^* means that the test is negative.

Bayes now says (and proves it mathematically) that:

$$P(A|E) = \frac{P(E|A)P(A)}{P(E|A)P(A) + P(E|A^*)P(A^*)}$$

or

$$P(A|E) = \frac{P(E|A)P(A)}{P(E)}$$

Table 2.1 Test results for 1,000 patients

	Patient ill	Patient not ill	Total
Test positive	1	20	21
Test negative	0	979	979
Total	1	999	1,000

since P(E), the fraction or probability of a negative test in the whole sample, is often all we know.

Now say we test 1,000 patients of which 1 patient is ill. For the test results, what this means is shown in Table 2.1. Although we have found the one ill patient, we have found another 20 patients who we think are ill but are not.

Say we believed the test was perfect for the ill patients (i.e. $P(E/A)=1$); we now have a new estimate that 5 per cent of the patients who test positive are ill, which is the right answer. We would also have got this answer if we had been frequentists, but in many cases the evidence is much less clear than in this example.

We can also see that if we believed *nobody* was ill, we would still believe nobody was ill after we found 20 per cent of the tests to be positive. This is an important result for the Bayesian estimates of accident probabilities in cases when one of the parties involved is of the opinion that the accident cannot happen. Even in the face of an accident, the non-believers will state that the accident will not happen again, and Bayesian statistics would support them (Unwin, 2003).

SYSTEMS AND INTERFACES

If we want to find causes and consequences, it is often said that we have to look at 'the system'. This leads to the obvious question: 'What is "the system"?'

Now we can have a very narrow look at it. The light system of Figure 2.2 could be considered a system. This limited view of what constitutes a system also leads to a limited view on potential causes or causal chains for light failure. Investigations of failures of the light bulb cannot be broadened into investigations into underlying causes such as the vibrations of the car on which the lamp unit is mounted; the car is not part of the system. Therefore, it is very useful to cast a wider net and include a larger section of the surroundings in the system.

We could go so far as to include the whole car in this case. Then we could, for instance, investigate whether the vibrations of the car are in any way related to the suspension or the tyres and whether we can do something to make the vibrations less severe, or design a light bulb that can withstand

Figure 2.3 The car–road–driver–light system

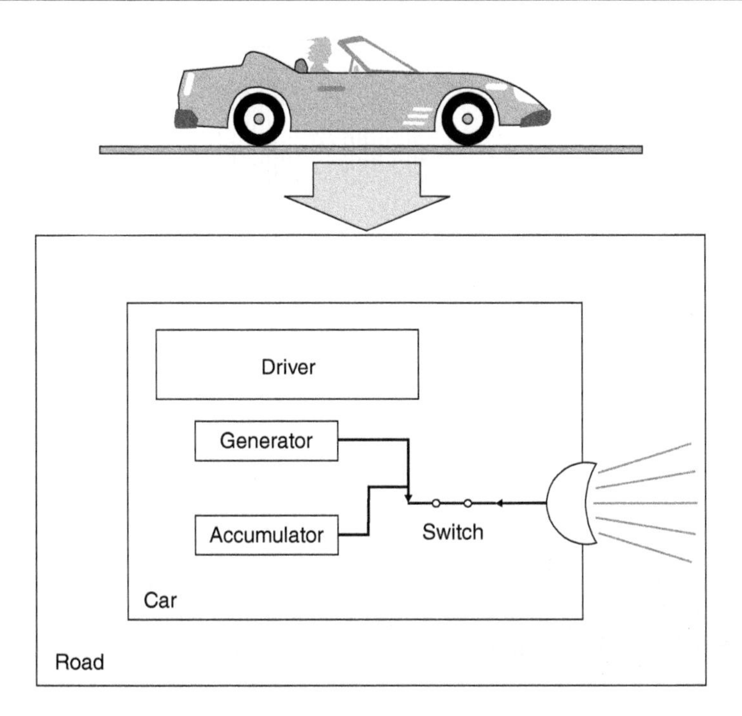

these vibrations. In that case we may discover that the condition of the road has something to do with the vibrations of the car and thus with the probability of a light bulb failure.

Having discovered that the road matters, we then discover that the speed matters as well. And speed? That is controlled by a driver. So now our system is already pretty large: it has a light in a car on a road driven by a driver (Figure 2.3).

Man, Technology, Environment

Industrial, technological and natural risks should be studied and controlled in the system consisting of Man, Technology and Environment (MTE). In this system, man – human beings – designs and operates technology, influences and changes his environment and at the same time is affected by the technology he designs and operates and by the environment that he changed (Kuhlman, 1981). The interactions between these three elements of the MTE system produce wealth and prosperity on the one hand and are the source of trouble and risk on the other. Man is part of this system and in part also the creator. Managing risk is all about controlling the MTE system, and risk science is all about understanding this system and the behaviour of it that results from human interventions.

Figure 2.4 The Man, Technology, Environment system

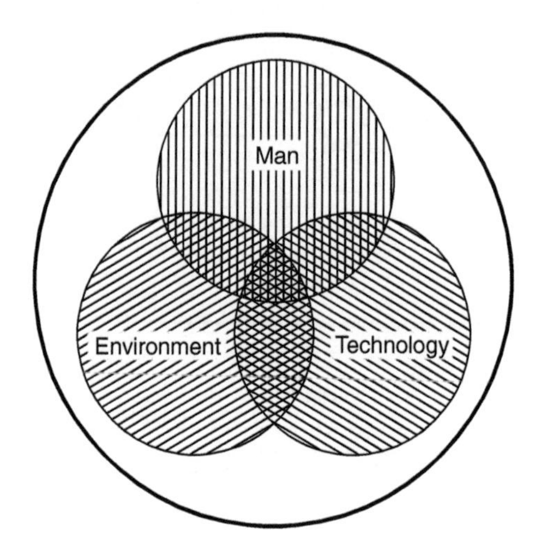

The most important step in understanding a system is grasping how it is put together. Figure 2.4 gives only a very superficial indication of the anatomy of the MTE system. It shows the areas of binary interaction between the three constituents and the area in the middle where all three constituents interact.

Accident Causation Metaphors

Three-dimensional picture

The idea of a three-dimensional picture is another form of model. In fact, the picture is just as two-dimensional as any other picture. This means that we cannot walk around or take another view from another angle of the object in the picture. Thus, things that are hidden 'behind' other things remain hidden.

There are various metaphors used to capture the essence of accident causation and the protection against harm. They include the hazard–barrier–target model, the Swiss cheese model and the bowtie model.

The concepts and metaphors that I shall describe originated from the practical question of where and how to prevent future accidents. For very rare accidents, the underlying causes are sometimes difficult to identify. Furthermore, every underlying cause has another, deeper, cause. Thus, the thinking about accidents evolves into ever deeper constructs and models. Nevertheless, any effort to construct a model describing chains of causality of events in a system must be based on the assumption that causality exists and that causality even in systems as complex as the aviation industry can be described. This unavoidably leads to the question of whether the work will ever be finished and whether there is a final cause to be found.

Several lines of discussion are continuing. Some of these are triggered by the perceived incomprehensibility of low probability–high consequence events, some of them by the notion that analysis of causality seems to have no end, and some by the more legalistic discussion on whether a probabilistic progression of a sequence of events should lead to a negation of the certainty of the cause after the fact.

The matter of causality is a highly philosophical question. I shall describe my position with respect to these questions briefly a little later, in order to justify the continuation of efforts towards those in the scientific community who have reached the point of seeing no further point in causal analysis and modelling.

The discussion about the infinity of the chain of causality is an old one and goes back to the Greek atomists around 400 BC (Russell, 1946). The *why* question in this context can have two meanings: 'to what purpose?' and 'with what cause?'. Both questions can only be answered within a bounded system, because they imply that there is something causing the system to exist.

A bounded system can show behaviour that the makers did not anticipate. In most cases the cause of this behaviour can be found as a combination of behaviours of parts of the system that the makers of the system did not consider. Projective analyses take time and effort, and efficiency requires these analyses to be limited. The fact that a behaviour was not anticipated does not imply that its anticipation was impossible, merely that it was deemed impractical.

Nevertheless, one could make the proposition that complex systems show emergent behaviour that not only is surprising, but could not be anticipated in principle. I share the position that this proposition is equal to proposing that the system is alive (Chalmers, 1996). And although human beings are part of the aviation system, I take the position that the aviation system is put together by humans and run by humans but is in itself inanimate (Arshinov and Fuchs, 2003).

As regards causality in the 'legal' sense, this is an issue that also plays a role in the discussion about flood defences. What causes a flood? Is it caused by high water or a low dyke? This is a question akin to asking what contribution the left hand makes to the noise heard when one claps one's hands. We consider the cause of the flood to be the combination of the height of water and the height of dyke, where the latter is lower than the former. Here, like Kant, we take cause to be a multi-attribute entity. More generally, a cause is the occurrence of a particular combination of the values of relevant parameters that give rise to an accident.

Figure 2.5 The domino metaphor

The Domino Model

The domino model first figured in the second edition of Heinrich's book on accident prevention. He depicted the events leading up to an accident as a chain of events. Each event in the sequence is the fall of another domino in a whole row of dominos. The only way of preventing the accident from happening is to remove a domino (Figure 2.5).

Although this model is considered to be a very linear way of looking at accident causation, it is in effect a very parallel way of looking at accident prevention. Any one of the dominos can be taken away to prevent an accident. This therefore is no different from raising one or more barriers against a cause progressing to an accident, as in the barrier model which we shall look at next.

The Hazard–Barrier–Target Model

The hazard–barrier–target model sees accidents as the result of a continuous threat, the hazard, on a target (Schupp *et al.*, 2004). This target is shielded from the hazard by a barrier. There may be one barrier but there

Figure 2.6 The hazard–barrier–target model

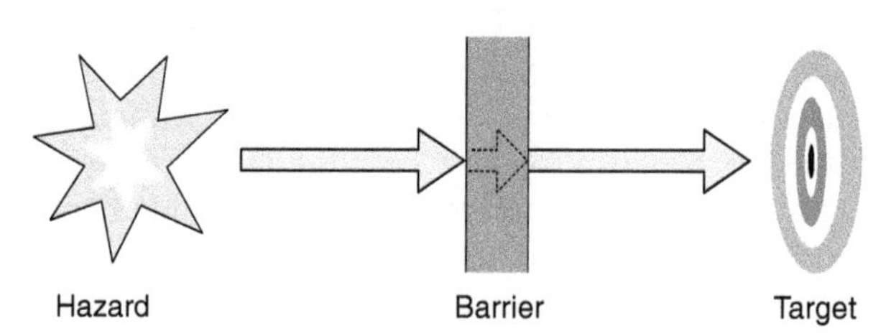

Figure 2.7 Pictorial model of road accident causation

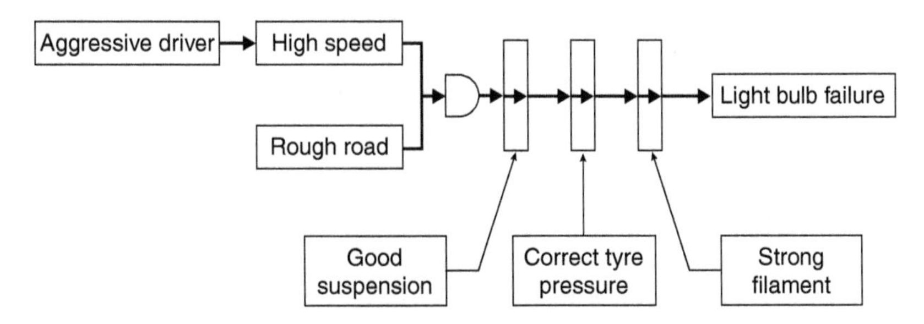

could also be several. These barriers need to be maintained. The maintenance of the barrier has to be secured by a – barrier – management system, also called safety management system or SMS. Barriers may be imperfect or absent as result of technical or human failure (Figure 2.6).

The Barrier Model

In our extended system the driver (Man) in combination with the road (Environment) is the cause of light failure when intermediate parts of the system – the suspension, the tyres and the strength of the bulb's filament (Technology) – do not prevent it. This allows us to make a pictorial model of accident causation for the accident: the failure of light, as shown in Figure 2.7.

The model says that the combination of high speed and rough road conditions will tend to cause a light bulb failure, but that good suspension, good tyres or a strong filament may prevent the failure from happening. It also says that the driver causes the high speed. The three preventive measures are also called defences or barriers.

This model is only one of the possible representations of the causal chain. It may be a worthwhile exercise for the reader to try to make alternative models. Doing so may lead to a better understanding of how this

Figure 2.8 Swiss cheese model

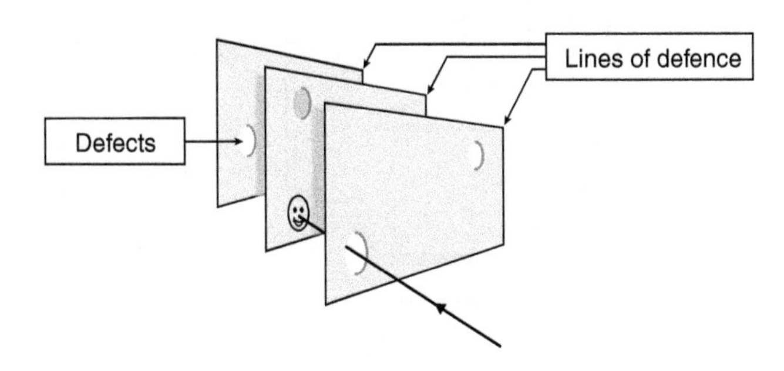

modelling work actually proceeds. In any case, more than one model is possible, and as long as the model serves its purposes, it is 'right'. We should be aware, though, that model and reality are never precisely the same.

Barrier-type models are also depicted in 'three-dimensional' diagrams, with the barriers represented as slices of cheese with holes in them. They are also called Swiss cheese models (Figure 2.8). The cause leads to the consequence if and when the holes in the slices are aligned and let the accident path pass.

Looking back at Murphy, we could also model human actions as finding the holes in the cheese, even if they do not line up. There are many ways of describing the same phenomenon. Each has its merits, each its pitfalls.

Defence in Depth

'Defence in depth' is terminology that originates in the nuclear power industry. In this industry there was on the one hand the demand to make the probability of the release of radioactive material extremely low. On the other hand, no single measure could possibly give the desired low probability. Therefore, nuclear power stations are designed with multiple barriers between the radioactive core and the surroundings. These barriers include multiple cooling systems and multiple gas-tight enclosures. Figure 2.9 is a schematic diagram of these features for a particular nuclear power plant that later would become notorious. The reactor for which the schematic is given is a so-called pressurized water reactor (PWR). A reactor of this type is used on Three Mile Island, near Harrisburg, Pennsylvania. In 1978 an accident occurred. First a safety valve remained open after there had been some overpressure in the reactor. As a consequence, radioactive water and steam leaked

Figure 2.9 Schematic of a nuclear power plant

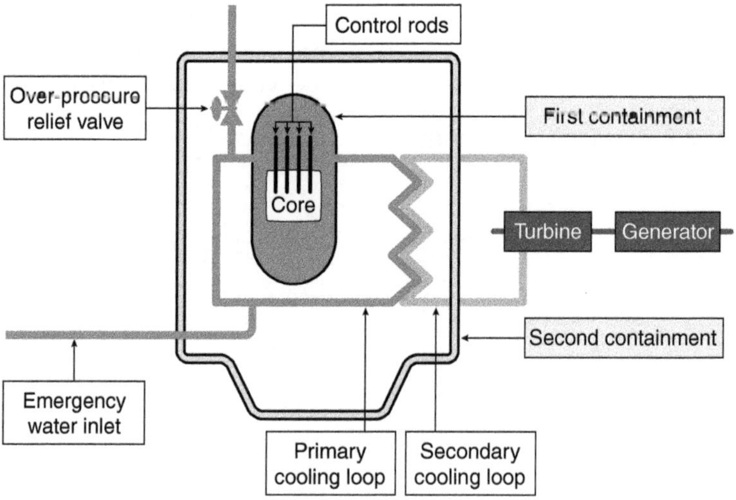

from the primary cooling system into the reactor building. The first barrier of the defence in depth structure was broken. This led to a sinking water level in the core, which the operators failed to correct (one more barrier breached). The core overheated and started to melt, increasing the amount of radiation. The leakage of water in the tank under the reactor vessel increased, until an automatic pump started to transfer water from the sump inside the building to a sump in a second building outside the primary containment building. Barrier number three broken. Luckily, this last building was airtight and no radiation leaked to the outside world. The operators succeeded in getting the situation under control before the core really melted.

The operators in Chernobyl were not so lucky. They had overridden most of the safeguards in the power plant because they wanted to perform an experiment on the reactor. Unfortunately, the experiment got out of hand. The reactor overheated and the core melted. This also released a lot of hydrogen in the reactor building, and it exploded and destroyed the containment.

Tripod

Tripod (Groeneweg, 1998) divides the various barriers into three kinds: faults, latent failures and preconditions. This way of distinguishing between different types of barriers is meant to dispense with the usual way of designating the cause of the accident as the last fault or mistake made, which usually also puts the blame on the operator. Groeneweg tries to emphasize what Reason (1990) also argues, which is that the operator is in many cases placed in a situation in which the accident has become unavoidable. This is illustrated in Figure 2.10.

In fact, this is a situation that arises in many more cases than is often assumed. A system is designed and design decisions have been made, which lead to a situation that entices, or even forces, the operator into performing an action in a way different from the one the designers had intended or anticipated. Although in hindsight this is often called wrong, to the operator the action seems completely normal. When this happens, the designers are usually long gone, so nobody notices, until the situation changes and some other condition occurs, which is abnormal. This then combines with the deviant action to create an accident.

Figure 2.10 Preconditions and latent failures

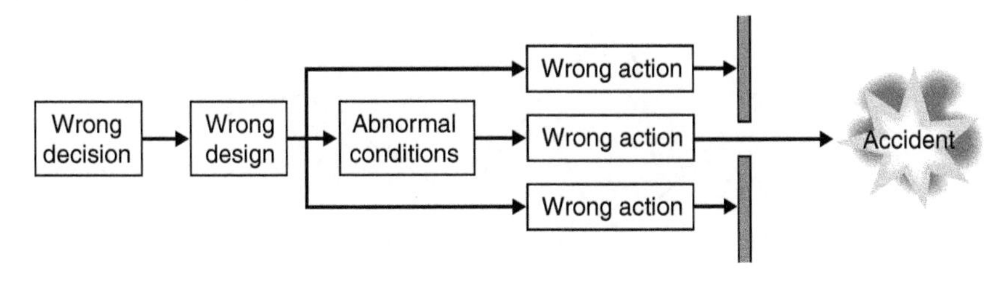

As an example, take the fire in Hoofddorp prison in 2005 (OVV, 2006). Eleven people died in this fire, all of them inmates of the prison. The immediate cause was established as a cigarette butt thrown away by an inmate. However, the fire resistance of the building was so poor, and the number of personnel and their training so inadequate, that a small fire, which under normal circumstances could have been extinguished without large consequences and certainly would not have spread all over the building, led to a major accident.

Functional Resonance

Modern high-hazard, low-risk systems, such as the aviation system, have developed such a high degree of technical and procedural protection that they are largely proof against single failures leading to a major accident, either human or mechanical (Amalberti, 2001). However, they are not proof against single failures that can trigger the realization of deeper and more widespread latent failures in the system. The aviation system is more likely to suffer 'organizational accidents' (Reason, 1990) in which latent failures arising at the managerial and organizational level combine adversely with local triggering events and with the errors of individuals at the execution level (Reason, 1997). This has led to another metaphor that is used by Hollnagel (2006), namely 'functional resonance'. Hollnagel tries to explain that reality is not a single chain of events, each event being the consequence of the previous one and the cause of the next. In functional resonance, all parameters of a system and all external influences vary over time.

Certain combinations of deviations from the mean, expected value bring a system outside the boundaries of control or of safety. The term 'resonance' is somewhat misleading, in the sense that the variations in the values of the parameters are not coupled by a mutual influence. 'Functional concert' would probably have been a better term (Figure 2.11).

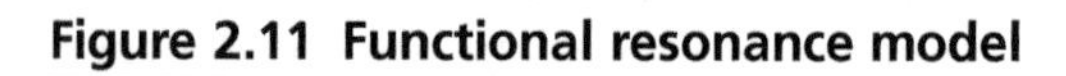

Figure 2.11 Functional resonance model

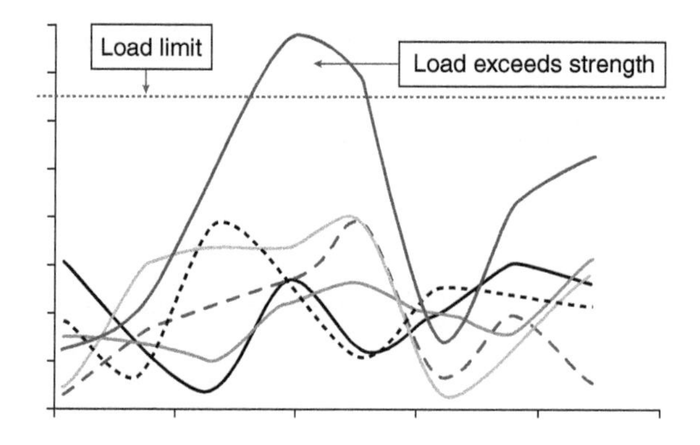

3 The systems approach

'Systems approach' sounds scientific. But what do we mean by it and why would we be taking this approach in safety science? In this chapter we shall see what systems and the systems approach are and how we can use the systems approach in dealing with risk.

People and machines seldom, if ever, operate in isolation. There is coherence and connection in the world. The world is a whole. So, we should really look at the world as a whole. But then, as mentioned earlier, there does not seem to be much going on that can really be understood. The world looks chaotic rather than ordered.

However, there do seem to be some laws – like 'action is reaction' and Murphy's law. In order to be able to understand how things work, how action creates an equal reaction, and what that reaction is likely to be, and how we fit in, and how we work ourselves, we need to analyse, to dissect and then put together again and integrate. Analysis and integration are the two complementary activities that we need to undertake to understand what is going on.

We cannot take everything apart simultaneously. We could not cope with such a task, and everything would stop working. So, we take bits and pieces that are large enough for us to get the picture and small enough for us to understand and make the work doable. What we take we call a system.

There are systems and there are systems.

THE SYSTEM

As we have seen before with the example of the lights of a car, what we call a system is almost always part of a larger system. We have already seen the general Man, Technology (the car), Environment (the road) system.

We could also have carved things up differently so that the resulting pieces would have been part of altogether different systems rather than car or engine.

In Figure 3.1 I depict two ways one could carve up transport. We could divide transport into the road transport system, the rail transport system, etc. We could also divide transport into transport companies that use various means of transport – vehicles – to transport goods. Most companies use more than one mode of transport. Even a company such as Air France/KLM

Figure 3.1 Vehicles are part of a transport company (system) and a transport network (system)

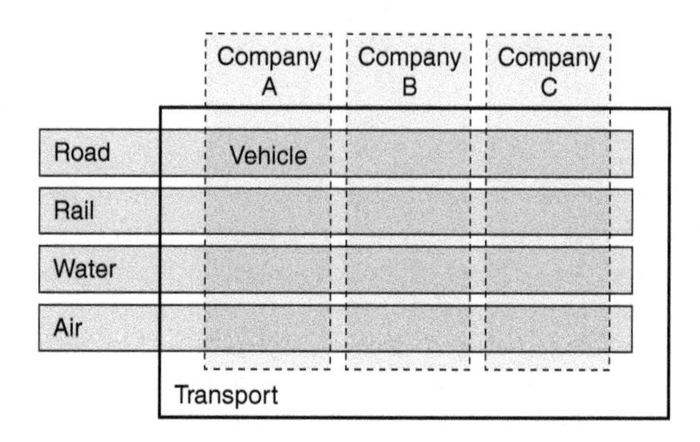

exploits both trains (high-speed rail) and road (to transport passengers and freight from and to airports). Other divisions are possible: vehicle manufacturing companies including their service network, fuel-producing companies. A vehicle thus can be part – that is, a subsystem – of many systems, which themselves also are connected, even when this connection is only this one vehicle that is part of all of them.

So, there is no one right way of defining a system and its boundaries. And when looking for faults and accidents, for causes and consequences, we often have to look at more than one such division at any one time. This is the price we have to pay for not being able to take a holistic view of the whole universe at once.

Where is the car?

We can try to list all the systems of which the lights of the car in our earlier example are part:

- the light system
- the electric system
- the car body
- the car
- the road transport system
- the car maintenance system.

And the list continues. It is a good exercise to try to make a long list of all the systems involved in having a car light work. This at once gives us insight into the complexity of society and into the complexity of detecting and fixing causes of failure – and thus how difficult it can be, and often is, to keep things working normally.

THE INTESTINES

We may try to understand a system by exposing it to outside influences and see what happens. We may also try to look at the inside, but doing so can sometimes be much harder than we expect. Sometimes it is impossible.

Black box

When we try to analyse a system, the first thing we can do is see how it reacts to what we do to it. What we do to it is called a stimulus and what happens is the response. In computer lingo the stimulus and response are also known as input and output respectively (Figure 3.2).

In many cases it is sufficient to know the behaviour of a black box. We often do not really care why what happens, happens. If we throw a switch (stimulus), the lights go on (response). If we press down on the accelerator (stimulus), our car goes faster (response).

In fact, most systems are black boxes to most of us, and we treat them like that. If a machine does not work, we kick it and see what happens. This was quite an effective way of getting a machine to work a century and a half ago. However, it is not a good strategy with modern electronics. But look around at how many people hit a computer when it does not work. Modern strategies are to click the left mouse button a few times, press the return button or press control alt delete. These are all black box techniques that serve us well. We do something to a system and see what happens, in the line of centuries of experimental investigation of the world around us. But sometimes ...

The hammer and the feather

The Italian scientist Galileo Galilei (1564–1642) first noted what would later be the foundation of Newton's theory of gravity. The acceleration of objects that fall to earth does not depend on their mass. All objects theoretically fall at the same (increasing) speed. This, however, presents a problem. Clearly, a feather falls more slowly than a hammer. So, there must be another component in the system besides the earth, the hammer and the feather. This component is the air. The resistance of the air slows the motion

Figure 3.2 Black box

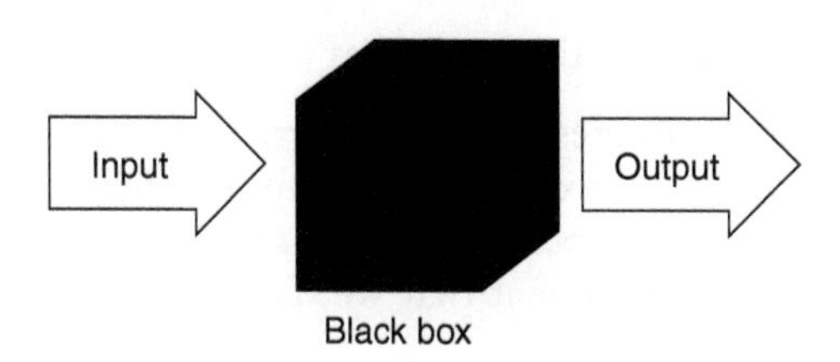

of objects through it. Since the forces on hammer and feather are different, the hammer will move through the air more quickly than the feather. If one takes objects of the same size and shape, on which air resistance is the same at the same speed, these objects will fall with equal velocity even if they have different weights. Galileo had to experiment using the tower at Pisa, and thus had to devise ways of getting around the problem of having air present. So, he made balls of the same size but of different weights and threw them from the tower – and observed that they hit the ground at the same time. In the age of space travel, an experiment can be done in the absence of air, and indeed one was done on the Moon during the Apollo 15 mission (see the box).

At the end of the Apollo 15 moon walk, Commander David Scott performed a live demonstration for the television cameras. He held out a geological hammer and a feather and dropped them at the same time. Because they were essentially in a vacuum, there was no air resistance and the feather fell at the same rate as the hammer, as Galileo had realized it would hundreds of years before; all objects released together fall at the same rate regardless of mass. Mission Controller Joe Allen described the demonstration in the 'Apollo 15 Preliminary Science Report':

> During the final minutes of the third extravehicular activity, a short demonstration experiment was conducted. A heavy object (a 1.32-kg aluminium geological hammer) and a light object (a 0.03-kg falcon feather) were released simultaneously from approximately the same height (approximately 1.6 m) and were allowed to fall to the surface. Within the accuracy of the simultaneous release, the objects were observed to undergo the same acceleration and strike the lunar surface simultaneously, which was a result predicted by well-established theory, but a result nonetheless reassuring considering both the number of viewers that witnessed the experiment and the fact that the homeward journey was based critically on the validity of the particular theory being tested.

Joe Allen, NASA SP-289, Apollo 15 Preliminary Science Report, Summary of Scientific Results, source: pp. 2–11

The important lessons from the above are that

- usually we become interested in the internal workings of what hitherto has been a black box if we obtain unexpected outcomes;
- if the outcome of an experiment on or in a system is unexpected, there usually is a component that we did not consider part of the system, but is part of it.

Understand the system

We have seen that a system reacts to inputs by giving outputs. We have also seen that what we define as being the system does not necessarily comprise everything. There may be external influences. like the air in Galileo's experiment. We may decide to take the source of these influences inside the system, making the system larger.

There may also be things that happen unexpectedly, because there is something going on inside the system that we do not understand or have not taken account of. This often happens when we have put the system together ourselves. Let us look at another accident.

TWA 800

> On July 17, 1996, about 20.31 eastern daylight time, Trans World Airlines, Inc. (TWA) flight 800, a Boeing 747–131, N93119, crashed in the Atlantic Ocean near East Moriches, New York. TWA flight 800 was operating under the provisions of 14 Code of Federal Regulations Part 121 as a scheduled international passenger flight from John F. Kennedy International Airport (JFK), New York, New York, to Charles DeGaulle International Airport, Paris, France. The flight departed JFK about 2019, with 2 pilots, 2 flight engineers, 14 flight attendants, and 212 passengers on board. All 230 people on board were killed, and the airplane was destroyed. Visual meteorological conditions prevailed for the flight, which operated on an instrument flight rules flight plan.
>
> The National Transportation Safety Board determines that the probable cause of the TWA flight 800 accident was an explosion of the center wing fuel tank (CWT), resulting from ignition of the flammable fuel/air mixture in the tank. The source of ignition energy for the explosion could not be determined with certainty, but, of the sources evaluated by the investigation, the most likely was a short circuit outside of the CWT that allowed excessive voltage to enter it through electrical wiring associated with the fuel quantity indication system.
>
> Contributing factors to the accident were the design and certification concept that fuel tank explosions could be prevented solely by precluding all ignition sources and the design and certification of the Boeing 747 with heat sources located beneath the CWT with no means to reduce the heat transferred into the CWT or to render the fuel vapor in the tank nonflammable.
> (National Transportation Safety Board, 2000)

From this summary of the investigation report by the US National Transportation Safety Board the cause of the accident looks fairly simple.

The vapour space of a fuel tank usually has a flammable or explosive mixture of air and fuel. There should not be a spark in such a space. So how could there be a spark there, then?

There has never been a definite conclusion, but the most likely cause is as follows. Wires of electrical circuits ran through the vapour space of these tanks. These wires had been there since the aircraft had been built, 24 years earlier. These wires were worn down as a result of friction as a consequence of the motions of the aircraft, which made the insulation thinner. The insulation prevents sparks from jumping from one wire to another if a high voltage occurs between two wires. Such a high voltage may be the result of 'auto-induction', which happens when equipment of high impedance is switched on or off. It can also happen as a result of a short circuit somewhere in the electrical system. It is believed that such a short circuit indeed occurred twice just before the aircraft blew up. When the aircraft was designed, its expected service life was some 20 years. In the course of the years following the construction of the aircraft, several modifications were made that extended the economic service life of these aircraft to over 30 years. The problem is the wiring. The insulation lasts some 15–20 years and then the wiring has to be replaced, including the wiring in the fuel tank, but these wires are extremely difficult to replace, so this was never done.

Interference

Two systems interfered in this accident: the electrical system outside the tank and the electrical system inside the tank – and the possibility of this interference had not been identified during design. Sometimes these interferences are much more abundant than one would assume. In many households the washing machine and the dishwasher are connected to the same electrical group. In the Netherlands such a group can have a maximum load of 3,300 W. These machines when running use a few hundred watts each except when the water is being heated; then they use a few thousand watts. Now the probability of that happening simultaneously for both pieces of equipment is small, but sometimes it does. Then the fuse blows and both machines stop working.

These influences between subsystems are hard to find during design, yet are often a cause of trouble. So, it is important to look for these unwanted connections and influences between systems and subsystems. Because we are not very good at getting our brain around a large and complex system, we have to analyse the behaviour of subsystems and their interactions in parts. Because we do it in parts we have to be careful that we do not miss something when we put the thing back together again.

Figure 3.3 Process, input, output, management

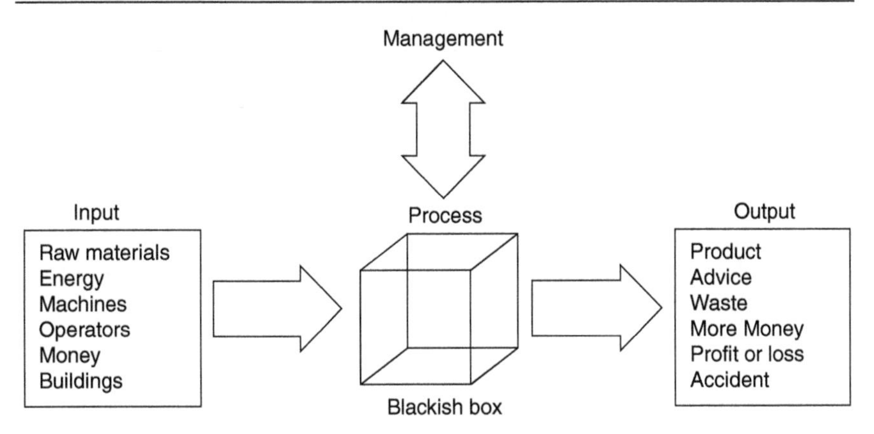

Systems Management

The black box of Figure 3.2 can now be extended to show the inputs and the outputs (Figure 3.3). In the system a process or processes take place that convert the inputs into outputs. If we put food, drink and a question into a student, the output may be the answer, which is what we want. The output can also be a happy student and no answer, which may not be what we want (unless we are the student). So, we have to manage this process. We could tell the student that he will get low grades if he does not provide us with the right answer in a reasonable period of time. Then the student may work on the answer or copy it from the internet (Figure 3.4).

Process management aims at controlling the behaviour of the system such that it will deliver the desired outputs for our inputs. Management is also a system. It is often subdivided into subsystems such as financial management, quality management, personnel management and safety

Figure 3.4 The student as a system

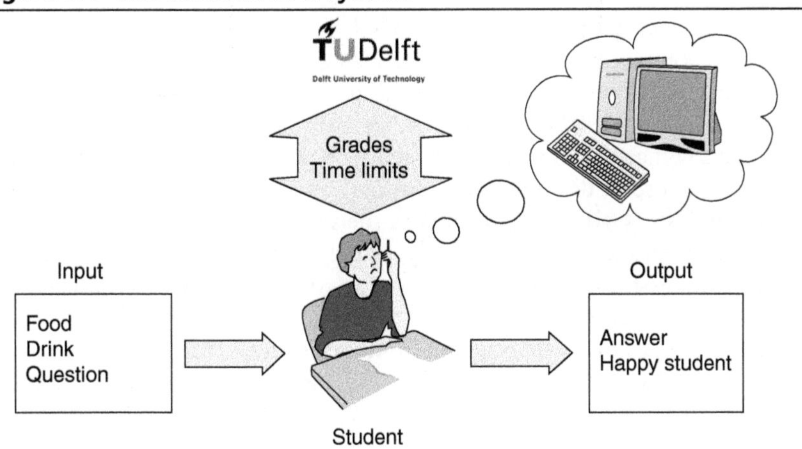

management. Often the management of personnel (humans), the environment and safety is considered as one cohering management task. This may be true, but it also suggests that such management is opposed to the real task of an organization: making money. However, in reality, risk management always serves the primary purposes of the system. Abnormal behaviour costs time and money, or defeats the purpose of the system.

A student who does not produce the right answers cannot graduate, which is the primary purpose of his or her being at the university to begin with. Therefore, spending money on having the student fed is a good investment for societies that need graduated people to support the wheels of the economy.

USING THE SYSTEM TO FIND SCENARIOS

When we have sufficient knowledge of the system, we can organize the way we identify the risks the system may pose, or the risks that the system is exposed to. This identification can then be followed and underpinned by fault tree analysis and quantification and the quantification of the magnitude and probability of the consequences.

There are a number of techniques that have become the standard toolkit of the risk analyst. For example, case studies can be used. Incidents and accidents of the past are studied to find out whether these provide indications of what could go wrong with the current system.

Another approach is to use checklists. A standard list of failures and accidents – often constructed on the basis of past experience – is checked to see whether similar incidents or accidents are possible in the current system. Methods based on historical evidence alone have the weakness of not being able to uncover new error failure modes that result from putting a new system together, even if this is done on the basis of existing components and technology.

Failure mode and effect analysis (FMEA) is a method in which for each and every component it is investigated what the modes of failure of this component could be and how these failures would propagate through the system (Dhillon, 1992). As this is an elaborate process, usually it is restricted to what are considered the main or most important failure modes, which in a sense undermines the purpose. Failure mode effect and consequence analysis (FMECA) is in essence the same method.

The hazard and operability study, originally developed by ICI, is one of the most successful methods. It uses a series of guidewords to challenge the imagination of the analyst. Although it is an elaborate method, the effort usually is worth it, as many defects and future problems can be dealt with at the design stage. Therefore, we shall discuss this method in the next paragraph.

Hazard and Operability study

The hazard and operability study (HAZOP; Chemical Industries Association, 1977) technique was originally designed by ICI and subsequently adopted by both the British and the worldwide chemical industry to get to grips with the complicated failure mechanisms of chemical plants. The original aim was not so much the safety of the workers or the surrounding population, but rather having the installation produce the chemicals it was designed to produce. Successful manufacturing of chemicals often depends on details, rather than the overall state of the installation or system.

It is often very difficult to make engineers think about what can go wrong in a system, just as it is difficult for most of us to consciously consider all the possible things that could adversely affect us, as doing so would make us very depressed. Therefore, the technique calls for a multidisciplinary team of experts – that is, not just the technicians who designed or built the system – to look at every vessel, tube, control valve, computer, etc. using guidewords.

The guidewords

Every parameter should be considered in conjunction with seven guidewords:

- not
- more
- less
- also
- partly
- reverse
- additional.

These guidewords are meant to fire the imagination of the team and help its members consider what would happen if the guideword in combination with the parameter were to occur. In the example of the headlights of a car that we used earlier, such a parameter could be the voltage of the battery.

Not, in relation to voltage, would mean that the battery was dead, which would be bad news – and not only for the headlights.

More voltage would obviously be a serious error, because it would mean that the wrong type of battery was installed in the car, and it would produce too high a voltage in the electrical system. Even if this did not cause the light bulb to burn out it would seriously shorten its life. It could also lead to excessively high currents and tensions in other parts of the system, leading to computer malfunctions and even a risk of fire.

Less voltage is less of a problem, as long is there is some, in so far the lights are concerned. However, it indicates loss of power and may create a problem when the driver tries to start the car.

Also (usually something else) is hard to envisage in the case of a battery, although the battery might leak acid, which would be detrimental to the metal (of which cars usually are built) if the leak continued for prolonged periods of time. And in the end the battery would stop functioning.

Partly in relation to voltage is in this case more or less equal to less.

Reverse is an interesting one. This could happen if the battery were connected wrongly, so that plus became minus. Connecting the battery wrongly usually ruins the car's electronic systems, but the lights would probably stay on.

Additional in this case is hard to image, but the reader is invited to have a go at it.

The causes

The next step is to figure out possible causes of the failures that have been identified. This is done only for those failures that are considered meaningful and relevant. In the case of the car battery, causes could be lack of liquid, low acid content, a short circuit in the battery, age, and human error in the installation. Obviously the first four causes could in turn also be the result of human error, but that would be the second layer and could be summarized by general lack of attention to the state and maintenance of the battery.

The probability

The next step is to estimate the probabilities. Sometimes this is done by rigorous fault tree analysis, but often an order of magnitude estimate is sufficient at this stage of the game. Detailed analyses can be done later, say in the process of a quantified risk analysis, when decisions are no longer obvious and it is necessary to weigh pros (risk reduction) and cons (costs).

The remedies

The next step is to define remedies. Experience has shown that in this step the main problem is to avoid the team finding staff training to be the only viable solution. As we have seen in the sections on human error, even after personnel have been given extensive training, human error can be provoked by complications in the situation, and by time pressure. This means that almost always the cause of the failure is a combination of technology and people. Consequently, almost always there is a technological alternative to training people. Given the fallibility of human beings, an attempt should always be made to design away the technical causes of the problem.

An example is the Ground Proximity Warning System (GPWS) installed in modern aircraft. These warning systems come in a number of generations, the most recent generation having the best approach warning and the least number of false alarms. Analysis of relevant crashes and crashes avoided just in time shows that the generation of GPWS system fitted in a plane is much more important than the training of the pilots in avoiding flying into the ground by accident – always provided, of course, that the pilot can fly a modern aircraft to begin with.

The result

The final result of the HAZOP is a large table with fault conditions, seriousness, probability estimates, remedies and usually an indication of the costs. This table is then put in front of the decision maker in charge, who has to decide on which remedies to implement and what residual risk to accept.

These risks include, as indicated earlier, potential loss of production or a lesser quality of product when the system malfunctions, even if there is no danger to the personnel or the surrounding population. For a full understanding of the risk, the results of the HAZOP need to be quantified and qualified, using fault trees and event trees and combining them into a quantified risk analysis (QRA).

The fault tree

Fault trees are designed to capture the causal chains that may lead to an accident and to depict and describe the logical relationships between the events leading to an accident. For the time being we shall use some basic symbols that indicate whether two or more causes have to occur simultaneously or whether one of them is sufficient.

Figure 3.5 Fault tree for light failure

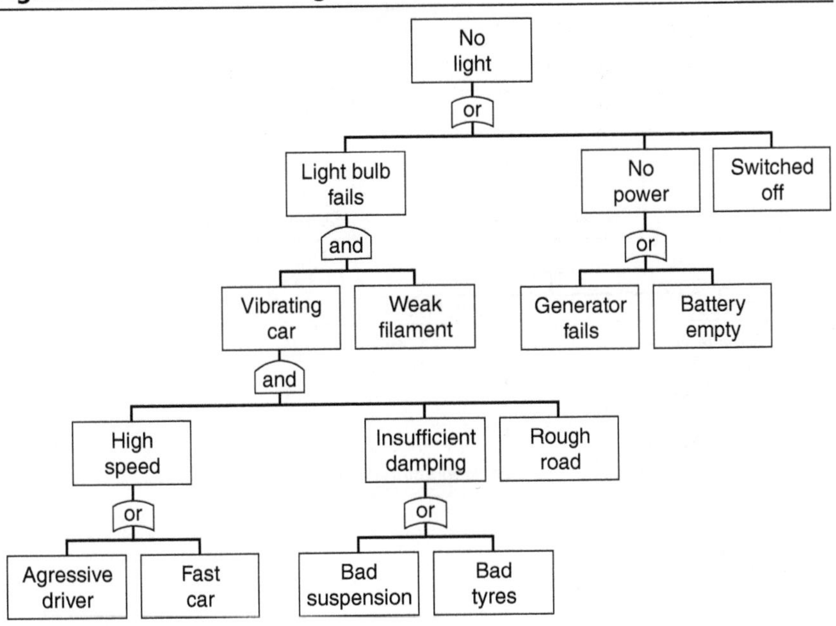

We can use the same example as before, the failing light bulb in our car. Remember that we have identified three reasons why the car light could fail: failure of the power supply, the switch being in the 'off' position or bulb failure.

We could further analyse the causes of filament failure into causes and measures that would prevent the causes from taking effect. In a decent HAZOP we would do this.

In Figure 3.5 all these elements are brought together into one fault tree. Apart from the boxes that contain the events we have already identified, such as a rough road or the battery being empty, two new symbols can be found in

Figure 3.6 Example event tree from the Causal model for Air Transport Safety (CATS)

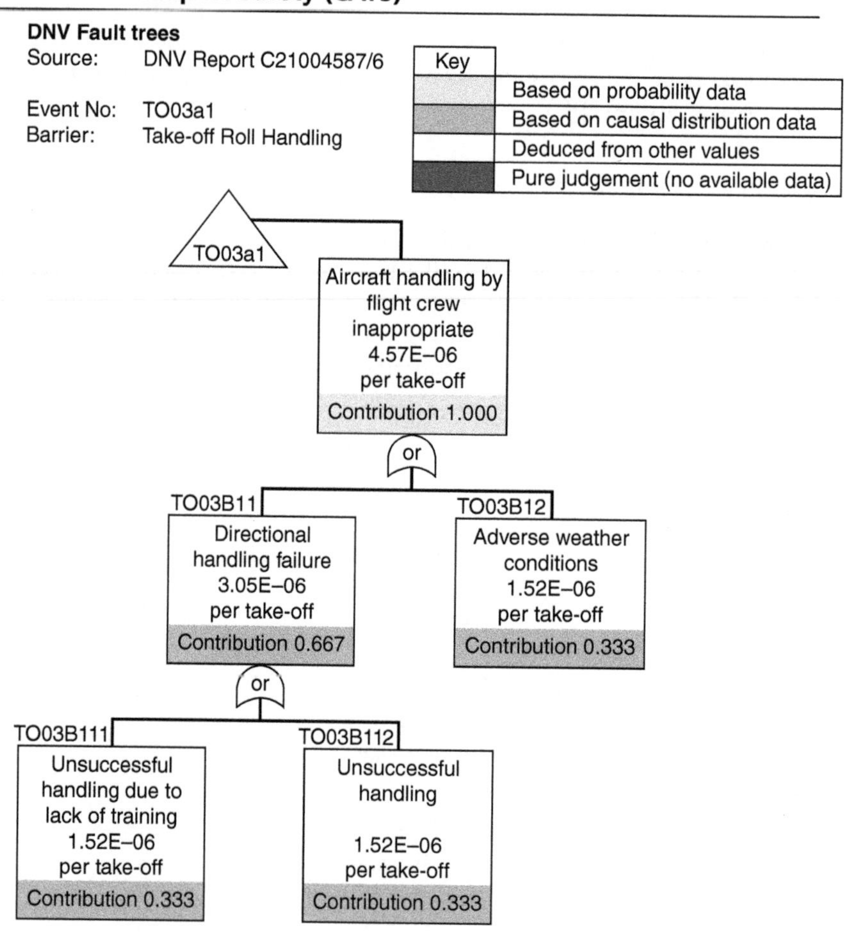

Source: DNV report C21004587/6

the figure. These symbols are on the paths from causes to consequences and are called **gates**. There are several types of gates in a fault tree. Figure 3.5 uses only two: the AND gate and the OR gate. When two or more events join in an AND gate, this means that BOTH have to happen for the consequence to occur. One of them alone is not sufficient. If the causes meet in an OR gate, however, only one need happen to cause the consequence.

A fault tree is not only a means for functional analysis; it is also used in making calculations. The probability of the top event can be calculated when the probabilities of all the base events are known. Base events are the events that do not have events or gates feeding into them. The mathematics of such a calculation is straightforward and was developed by people such as Arnaud and Pascal, albeit for gambling purposes initially. An example of a quantified fault tree from the Causal model for Air Transport Safety (CATS) is given in Figure 3.6 (Ale, 2006).

Quantification of fault trees

In a fault tree a whole series of symbols are used to represent the various events and the logic in the tree (Vesely, 1981). They include the symbols for the AND and the OR gates that were used in the trees earlier.

- An AND gate means that all the incoming events have to be true for the outgoing event to be true.
- An OR gate means that one of the incoming events has to be true for the outgoing event to be true.
- An exclusive OR means that the outgoing event is true when ONE of the incoming events is true, but not when more than one of the incoming events is true.

For the logic in the tree the following laws apply:

Commutative law:	X or Y = Y or X and Y = Y and X
Associative law:	X or (Y or Z) = (X or Y) or Z
	X and (Y and Z) = (X and Y) and Z
Distributive law:	X or (Y and Z) = (X or Y) and (X and Z)
	X and (Y or Z) = (X and Y) or (X and Z)
Idempotent law:	X and X = X
	X or X = X

The probabilities that the outgoing event is true can now be calculated using the following formulae

AND port	P(A and B) = P(A) . P(B)
OR port	P(A or B) = P(A) + P(B) – P(A) . P(B)
Exclusive OR port	P(A or B) = P(A) + P(B) – 2 . P(A) . P(B)
Mutually Exclusive OR port	P(A or B) = P(A) + P(B)

Figure 3.7 Some symbols used in the drawing of fault trees

Symbol	Name	Description
output / or / input	**OR gate**	The output occurs if one or more of the inputs to the gate occur.
output / and / input	**AND gate**	The output occurs if one or more of the inputs to the gate occur simultaneously.
	Basic event	A basic event represents a basic fault that requires no further development into even more basic events.
	Intermediate event	The rectangular shape is used to put descriptions of events that occur as output of the gates.
	House event	The house shape is used to represent a condition that is assumed to exist, such as a boundary condition.
	Underdeveloped event	A diamond is used to represent events of which the further development into basic events is not performed for reasons such as insufficient information, assumed insignificance of the event or reaching of the system boundary.
out / in	**Transfer symbols**	These symbols are used when the tree spans several pages. They are labelled such that the transfer points between pages can be identified.

With these formulae it can easily be seen that the result of the fault tree in Figure 3.8 is that the probability of the top event equals 2.2 x 10^{-5}.

However, one should realize that this result is true only when the base events are independent. When B and D in fact are the same event, and B (or D) occurs, the output of both the OR gates becomes true simultaneously and thus so too does the output of the AND gate. In that case the probability of the top event becomes 1.01 x 10^{-3}. In more complicated fault trees, handling of the common cause or common mode events is extremely difficult (Mosleh *et al.*, 1988).

Limitations of fault trees

Apart from the problem of common mode and common cause failures, fault trees have some other limitations. Fault trees are logic trees. The logic is called 'instantaneous'. There is no time element involved in a fault tree. This means that once the conditions are fulfilled, the consequence arrives.

This is often not really what happens in practice. Even with bad tyres and a weak filament the lights do not go out immediately. To model such a situation correctly, use needs to be made of other similar or derived techniques involving the calculation of probability, and the use of time delay. Examples

Figure 3.8 Simple fault tree

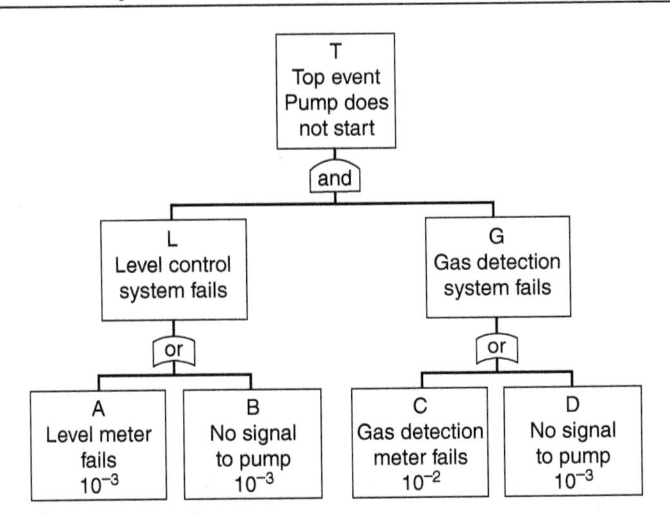

are event diagrams, Bayesian belief nets and Petri nets. These we shall consider later. For now we shall concentrate on the basics.

Before we can build an integral model we need one more component: the consequence tree.

The consequence tree

Until now we have looked only at cause–consequence relationships to find causes for a known or assumed event, which we called the consequence. We can also look downstream of the causal chain and see what a certain event may cause.

This is often a necessary exercise. We want to know whether small disturbances can have large consequences. We have considered how a butterfly can cause a thunderstorm and how a valve's failure to close almost destroyed a nuclear power plant. How a small incident can lead to a large disaster can also be seen from the following text by Prof. dr W.A. Wagenaar in the NRC (Nieuwe Rotterdamsche Courant) of 1982:

> Imagine an almost new cruise ship with a length of 120 metres. It can carry 425 passengers and 205 crew. The vessel complies with all national and international safety laws and has automatically operating watertight doors. Would you, reader, be able to sink this ship without the use of any tool and while the crew is continually trying to save the ship? Even Hitchcock would have had trouble inventing a credible scenario for this accident. Yet it is not difficult at all. Only one bolt has to fail, as we saw in 1980 when the Dutch passenger ship Prinsendam sank 120 miles west of Alaska. The cause was a long chain of human error that started when a bolt called a

'banjo bolt' in the low-pressure fuel supply broke. By concerted teamwork the crew subsequently found one of the few possible ways of sinking this ship seven days later. A full description of the disaster can be found in a 190-page report by the ship's council, which was published in the *Staatscourant* of 26 February 1982.

What, then, could be the consequence of the failure of a light bulb in my car? The reader could try to make a list. Here are a few possibilities.

I could drive off the road and hit a tree. Or I could drive into a ditch. I could hit a tree, thereby destroying the car. I could end up injured or dead. The car could also catch fire, with me in it. In Figure 3.9 the consequence tree has been expanded a little, but the reader should easily be able to think of further possible ways of developing this tree to the right.

How far we need to develop such a tree depends on its purpose and the consequences that we wish to consider. Often these consequences are material damage (money), injury or death. An example from the CATS report is given in Figure 3.10. This diagram, which in the CATS report is also called an event sequence diagram (ESD), shows the logic of the main events leading to a collision of an aircraft with the ground due to a so-called controlled flight into the ground, which means that the pilots steer an otherwise completely normal aircraft into the ground.

Whereas in a fault tree each node can have only two states, true or false, the nodes in a consequence tree can have multiple states. This is especially useful when the outcome of an event is continuous.

Consider, for instance, the failure of a pressure vessel. In a two-state world only two outcomes can be defined: the vessel is whole or the contents

Figure 3.9 Part of the possible consequence tree resulting from the failure of a car light

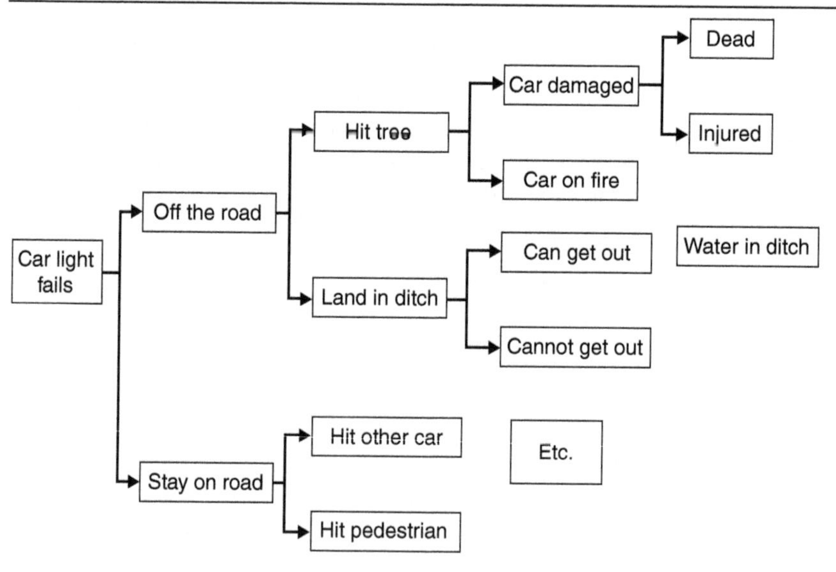

Figure 3.10 Consequence tree from the CATS study for a controlled flight into the ground accident

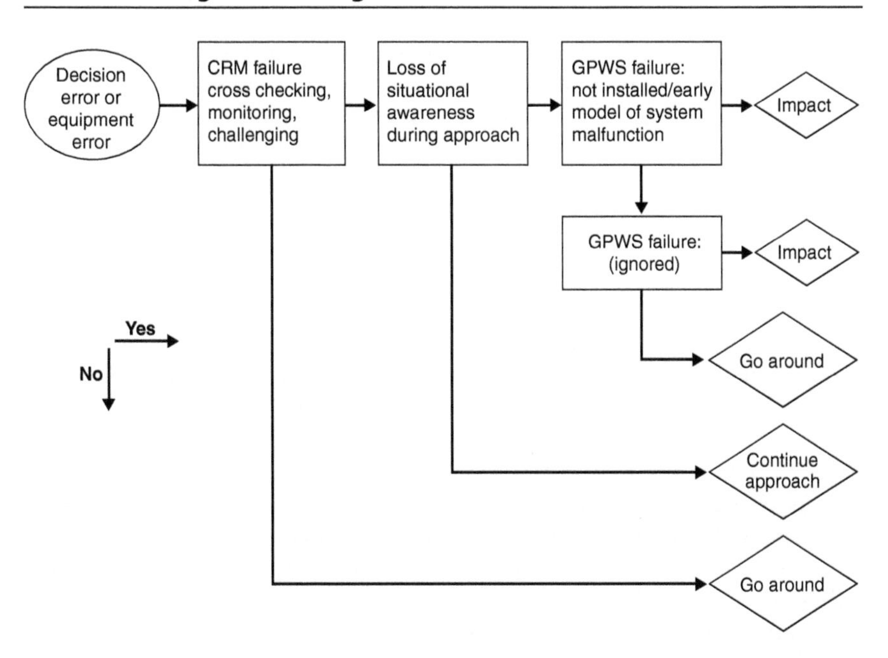

Figure 3.11 Multistate logic in a consequence tree

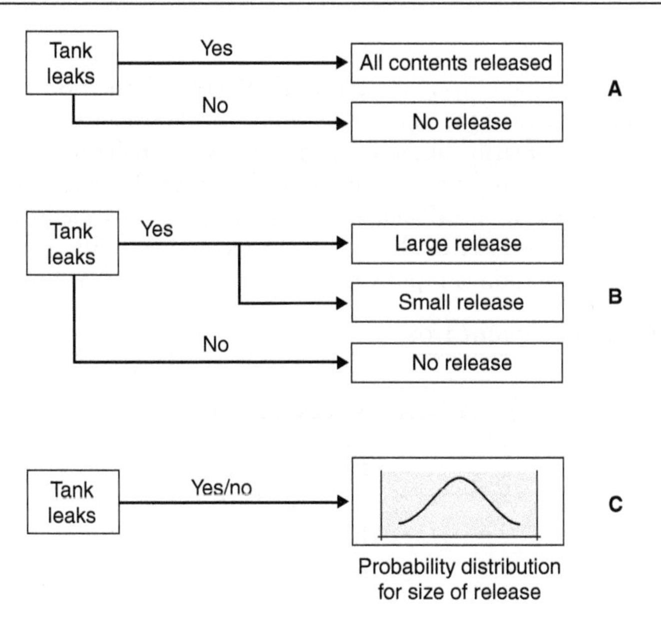

are released completely (Figure 3.11A). However, it is also possible to define several potential sizes of releases such as large and small (Figure 3.11B), or small, medium, large. One could also define the outcome as a distribution of release sizes, but this is beyond the capabilities of a tree model. For such refinement Bayesian belief nets are used, as we shall see later.

Although using multi-state logic is possible with an event tree, in practice it is much more convenient to define the tree in two-state logic only. This makes quantification much easier. However, sometimes multi-state logic is unavoidable (Papazoglou and Ale, 2007).

Quantification of a consequence tree

The probability of occurrence of the consequences can be calculated when the conditional probabilities of exiting each of the boxes (or events) in the tree are known. A simple example is given in Figure 3.12. Note that the sum of the exiting probabilities should always be 1. It can easily be seen that P(F|A) = 0.245. The reader is invited to work out P(D|A) and P(E|A).

As well as the probability, the consequences need to be assessed too. For instance, scenarios developing from a loss of containment accident can be evaluated, resulting in a distribution of concentration of chemicals, a distribution of heat radiation or a distribution of pressure waves – whatever is appropriate given the material released and the characteristics of the release event. An essential piece of information at this stage is the dose–effect relationship. This is the relationship between the exposure to a potentially harmful agent and the actual damage done to the subject of concern.

Target information

To calculate the damage from the physical effects, information is needed about the vulnerability of the objects for which the damage has to be calculated. These are primarily human beings. In some analyses, damage to structures like houses and offices has been quantified. Damage to surface water is increasingly often calculated. The damage to ecosystems is rarely included in the quantitative analysis of a chemical plant, although interest in this kind of calculation is rapidly growing. In the event that the target is

Figure 3.12 Simple event tree

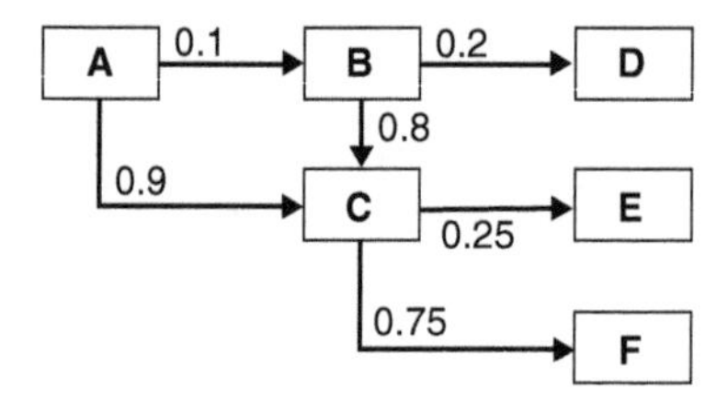

a human, the effects of heat radiation and exposure to toxic materials by inhalation or through the skin should be quantified. Relationships between the dose and the effect are taken from the literature. Information about the vulnerability of people to various physical effects was first systematically assembled by Eisenberg and colleagues in the 'vulnerability model' developed for the US Coast Guard in 1975 (Eisenberg *et al*., 1975). Based on this work, extensive research has been done and is ongoing into the various relevant dose–response relationships. The results are given in a guidance document called the Green Book (CPR, 1990). The Green Book is the result of almost two decades of close cooperation between various institutes: government, private and industry.

The dose–response relationships can take two forms:

- A threshold value below which the damage is nil and above which the damage is total. This is for instance the case for explosion effects. When people are subjected to an overpressure of more than a certain value (typically 0.3 bar overpressure), they are considered dead, otherwise it is considered that they will survive. This absolute division between yes and no is an approximation of reality. Although a person who is crushed under a 50-ton container will not survive, there are many instances where a supposedly sublethal effect kills a person and, conversely, people subjected to what generally is accepted as a lethal effect sometimes survive. So, a probabilistic relationship is usually better. But often there are insufficient data on human exposure.

Figure 3.13 Risk calculation: scheme in principle

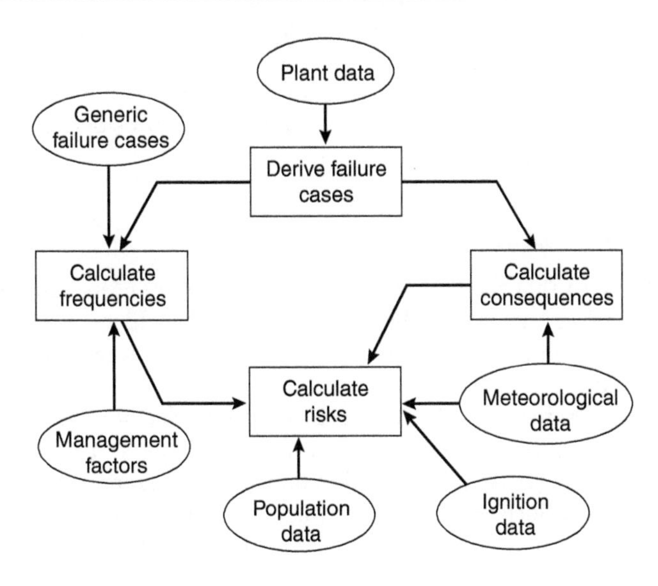

- A relationship between the dose or the exposure and the fraction of people who are expected to be killed or are likely to suffer some other predefined – harmful – effect. This is the case, for instance, when one is calculating the damage of a toxic cloud or a jet fire. These relationships are often referred to as *probit equations*, since they employ a mathematical entity called the probit, which converts the generally S-shaped relationship between dose and fraction of people into a straight line. With this kind of relationship the physical effects can be converted into damage. A useful way of expressing this damage is the fraction of population in a certain area that will be killed or the chance that an individual in a certain place will be killed, or will suffer some other effect, given that the initial event has happened (Figure 3.13). Similar relationships are available for material damage, although they are usually less probabilistic in nature (Ale and Uitdehaag, 1999). Nevertheless, some recent work does address the probability aspect for material damage.

Methodology

An essential piece of information is the presence and distribution of the targets, whether these are people, buildings, infrastructure, ecosystems or any other subject of concern, as well as the associated dose–effect relationships just discussed. Just as for the hazard analysis and the event scenario assessment, the Purple Book gives detailed guidance of how to do this, and there are many computer tools that will perform the necessary operations. However, probabilistic relationships between potential and harm, such as the probit relation, associate a probability with the level of harm. This probability has to be taken into account when addressing the likelihood of damage later in the process.

Other factors such as the probability of weather and wind, which may have a significant influence on the damage, are considered when building the final risk picture.

Bow tie diagram

We have seen that the fault tree ends in a single top event, and that this tree has the causes and causal chains leading to the accident. We have also seen that the consequence tree has all sort of consequences that follow from a single event.

It is therefore an obvious idea to couple the two into a diagram as depicted in Figure 3.14, which by reason of its shape is called the bow tie diagram. As far is known, this was first done by the hazard analysis department of ICI (1979). Such a diagram has the causes on the right, the consequences on the left and barriers wherever they are deemed possible or thinkable.

Figure 3.14 The basic bow tie diagram

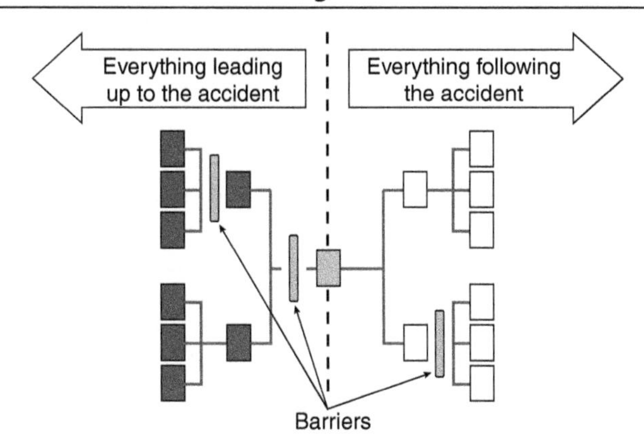

From the original idea many variants have been derived (Groeneweg, 1998; Visser, 1998). The latest development is that of the bow tie as an event sequence logical diagram, which can be used for the quantification of occupational risk (Ale, 2006). Here the centre event again is an accident. As an example, Figure 3.15 depicts the bow tie for falls from ladders. The fall is represented by the circle in the middle. Potential causes, such as inappropriate positioning of the ladder or instability of the user, are represented by the boxes on the left of the diagram. On the right-hand side one can find barriers or remedial actions and other factors that contribute to the consequences or may mitigate them – in this case, the height of the fall, the type of surface and the swiftness of medical assistance, for instance.

Bayesian belief nets

The diagrams used so far have all been trees. This means that from a single event, such as the top event of the fault tree, the starting event of the event tree or the centre event of the bow tie, the trees develop by dividing each event into parent or child events at gates. Once divided, branches cannot reunite.

The trees also have a direction. Fault trees go from base events to top event, consequence trees from starting event to ultimate consequences, and bow ties from causes to consequences. Diagrams with a direction and no cyclic paths are called directional acyclic graphs (DAGs; Bedford and Cooke, 2001). If the events represented by the nodes have a probability associated with them that depends on the states of parent nodes, this can be translated into the probability that the whole network is in a certain state. Such a probabilistic net is also called a Bayesian belief net (BBN). Frequentists would probably have called them state probability nets.

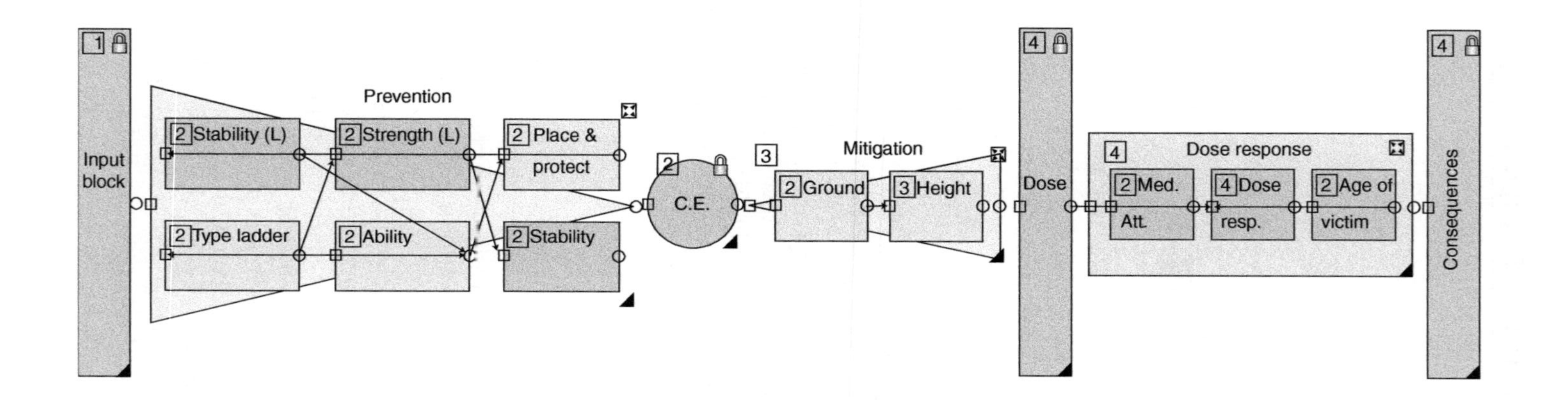

Source: Ale (2006)

Figure 3.15 Bow tie diagram for falls from ladders

Everything that can be done with fault trees and event trees can also be done with BBNs, but not all that can be done with a BBN can be done with fault and event trees. The handling of probabilistic relationships and distributed outcomes can only be done through BBNs.

The simplest form of BBN has only two nodes, as in Figure 3.15. Implicit in this diagram for each node is that the event that the node relates to is in the true state, or that the event meant has indeed happened. The probability that event x happens is $p(x)$ and the probability that event y happens is $p(y)$. The probability that y is true depends on the state of x. The probability that y happens when x is true is $p(y|x)$, which makes the probability that x and y happen simultaneously

$P(x,y) = p(x)p(y|x)$

We note that $p(x,y)$ is not necessarily equal to $p(y)$, as there might also be a probability that y happens when x has not happened. This cannot be the case in a fault tree or an event tree with the same structure as in Figure 3.16. BBNs do not need to be trees. They can have branches join later, as long as no cycle is introduced. This allows much more complicated models to be built and quantified, such as the model for the failure to execute a missed approach procedure in the CATS model (Figure 3.17). These nets allow studies into the probabilities of certain events or combinations of events given that other events have happened or are true. For instance, in the model shown in Figure 3.17, one could set the wind speed to a certain value (Ale *et al.*, 2008).

Software is available that enables the probabilities or probability distributions of all other nodes to be worked out. This possibility means that one can investigate not only whether the expectation of the probability of a certain event can be influenced, and by how much, but also how the probability distribution that results from the variability of the many events in the model can be influenced and changed. In the study of systems, events are always states of the system as a whole or states of the various components. BBNs are therefore a powerful tool in the study of failure probabilities of systems, whether you are indeed a Bayesian, or a frequentist after all.

Figure 3.16 The simplest belief net

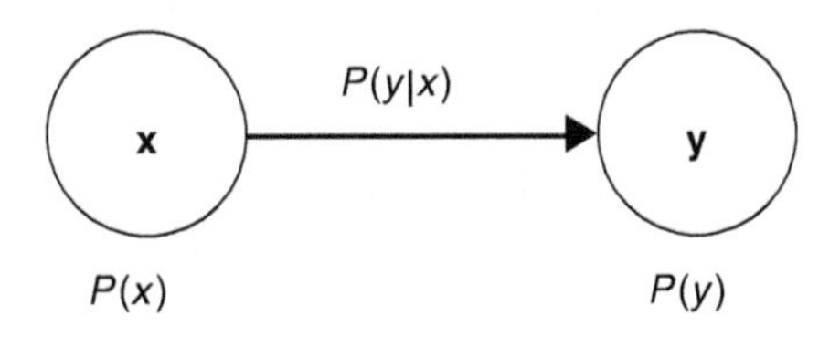

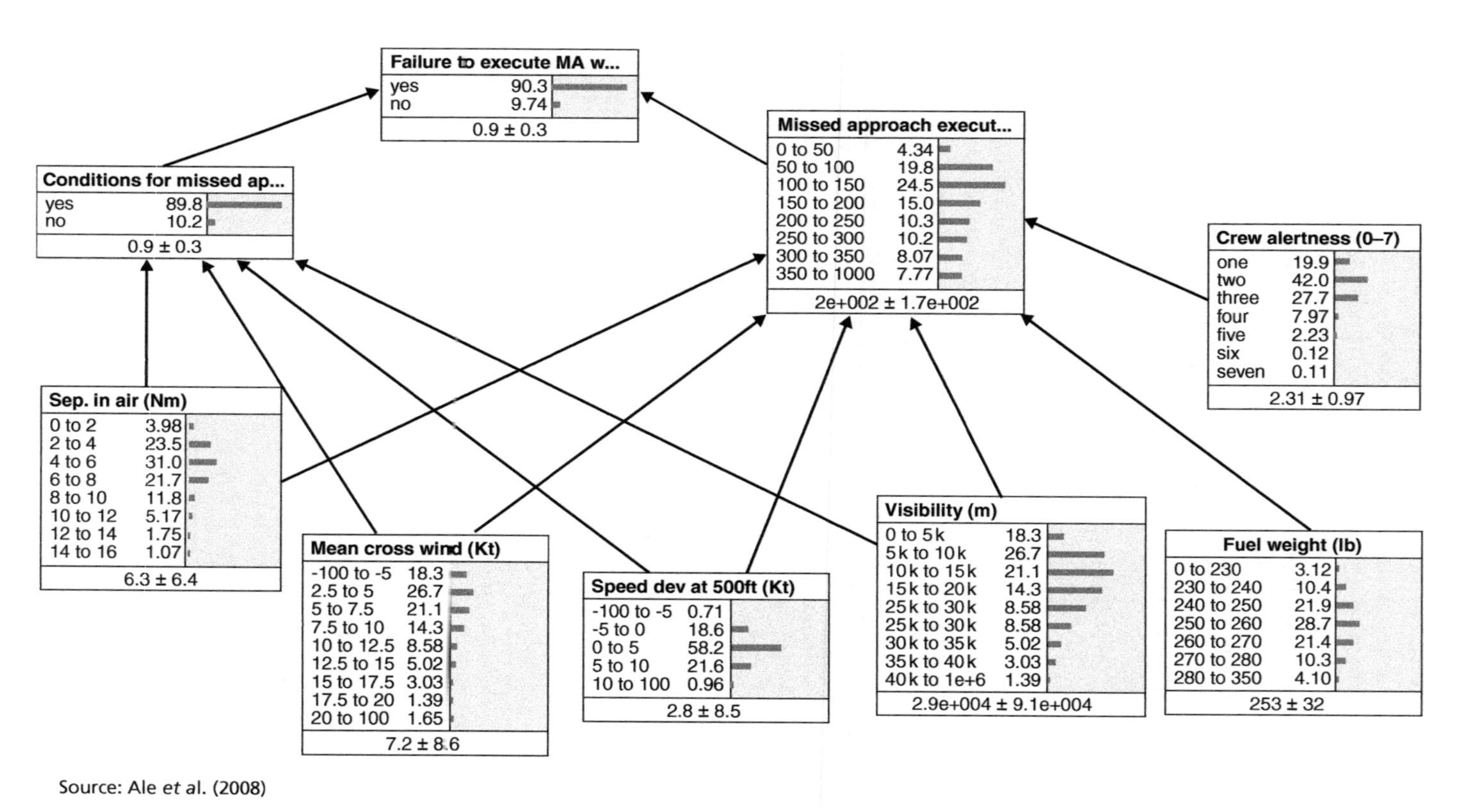

Source: Ale *et al.* (2008)

Figure 3.17 Bayesian belief net (BBN) for missed-approach execution from CATS

LIFE CYCLE

We have seen that the behaviour of a system cannot be understood without taking into account the systems with which it is connected and with the surroundings or environment in which it operates. We have seen that we can determine how these influences can be taken into account when analysing the effect of a stimulus on the system. If this stimulus is a fault, or faulty behaviour, or simply a deviation from normal, the system may fail.

We also have to take into account that a system does not stay the same over the whole of its life.

Time erodes all.

So, systems age over time. Age can change strength, but also, for instance, the conductivity of wires, the resistance of resistors and the capacity of capacitors. In general it can be observed that the probability of failure of a system depends on its age. When a system is young, the probability of failure is high because of initial failures. Once the system has settled down in its operation, the failure rate is constant over a long period of time. By the time the system comes to its technical end, the failure rates go up again (Figure 3.18). Failures during the early stages of operation are usually caused by design errors and errors in construction. The later errors are due to wear and tear.

The early errors can be found by checking the design, perform hazard and operability studies and by monitoring the construction, especially for conformity to the design. Failure to do so has led to many accidents and incidents such as the subsidence of a new building in Amsterdam. In this building the reinforcement in the concrete differed from what was specified in the plans in many places. In some places the difference actually made the building stronger, but in many places it made it weaker than it had been designed to be. Shortly after the building was put into service, cracks occurred. The building had to be evacuated and repairs had to be carried out to make the building safe.

The later failures can be prevented by preventive maintenance. We shall see how this influences risk management and safety.

Figure 3.18 The probability of failure depends on age: the bathtub curve

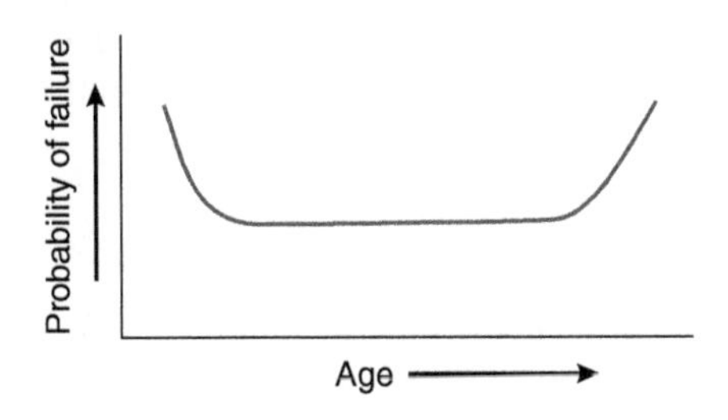

MITIGATION

Most of what we have examined up to this point has had to do with prevention. We have discussed how barriers could be placed between cause and accident or between cause and consequence. Most of these barriers were located on the left-hand side of the bow tie, before the accident. Some measures were discussed that form a modelling point of view could be placed on the right-hand side, such as shielding, and safety distances.

The last resort is to mitigation. Mitigation aims at making the consequences as minor as possible. Among these are evacuation, medical attention and financial compensation. A well-organized emergency response can contribute a great deal to the mitigation of the consequences and the trust of the population (Reniers *et al.*, 2007, 2008). On the other hand, a disorganized response may make things worse and will anger the population (Oosting, 2001).

Trust in the authorities is one of the important determinators in risk perception. Therefore, financial compensation should be generous and swift if the authorities do not want to lose people's confidence. When people had to be evacuated after cracks developed in their apartment building in Amsterdam, the city compensated them for the expense they incurred (OBL, 2007). However, the tax office subsequently removed part of that compensation, showing little understanding of the need to restore trust in the authorities, and little compassion for the victims.

Especially when the risks of large-scale disasters, such as flooding or an imminent major accidental explosion, are under discussion, one of the potential measures that can be taken is to prepare for and execute evacuation. Many years of research have been undertaken to assess the possible effect of evacuation.

Evacuation

The fate of the population depends strongly on their own ability to find rescue and – usually later in the course of an accident – on the capabilities of the rescue services to move them to a safe place and treat their injuries. This in turn depends in the first instance on how high the concentration is going to be, how fast significant concentration levels are reached and how long these last.

The public may be instructed to behave in certain ways, for instance to create a safe place in their home and to shelter indoors, or to evacuate in a certain direction (crosswind). Giving such instructions may need preparation, and the decision to actually give them may take time.

Dose–response

The exposure to chemicals influences people's ability to take evasive action when needed. Since 1986, studies have been undertaken (Bellamy *et al.*, 1987) on the factors affecting evacuation and crisis response, especially in relation to incorporating an understanding of human behaviour into crisis and emergency planning, and in particular into risk assessment for toxic and radioactive releases and for fires. Since the nature of the human response by those under potential threat and the time it takes to reach safety are crucial factors in both crisis planning and modelling, these factors and the parameters affecting them have been investigated in some detail. The responses of crisis managers and the human resources used in crisis management are also key components in the modelling.

In the 1980s, Bellamy began collecting data on evacuation behaviour in response to different kinds of disasters, and these data provided a basis for developing a model for escape and evacuation in the event of toxic releases (Bellamy, 1986; Bellamy *et al.*, 1987) which was later developed as a module in the SAFETI risk assessment software. This work also provided a good foundation for work undertaken by HM Nuclear Installations Inspectorate, who were interested in evacuation around nuclear power plants and the implications for emergency planning (Bellamy, 1987, 1989a, b; Bellamy and Harrison, 1988; Harrison and Bellamy, 1988). Furthermore, this work also provided a good basis for considering evacuation from fire in buildings and from offshore platforms. In work carried out for the United Kingdom's Building Research Establishment Fire Research Station the subject of warning systems was tackled in relation to the lag time between the realization of the threat and people's response to it (Bellamy *et al.*, 1988; Bellamy, 1989c; Bellamy and Geyer, 1990), showing how different properties of warning systems can affect speed of response. Work carried out on offshore systems showed how scenarios developed for risk assessment purposes can be used to help those concerned understand the implications for emergency response planning (Cox and Bellamy, 1992), and similarly how the nature of the human response can be used for developing scenarios (Bellamy, 1993; Bellamy and Brabazon, 1993).

It is also important to understand the dynamics of crisis management itself, and this can also be modelled (Drieu, 1995) to identify effective (and ineffective) interactions with the mitigation of the threat and the bringing of people to safety.

All in all, these studies have shown that an understanding of human behaviour under threat or potential threat or in crisis management is fundamental to risk assessments that are used in crisis decision making and other risk reduction activities. There is an increasing need for crisis management to be proactive in this respect (Bellamy, 2000), and hence to ensure that the demands on crisis management are reduced through elimination of scenarios that are unmanageable or that exceed the available management resources. This is illustrated in Figure 3.19, where curves a–e represent scenario development (harm over time) in the absence of intervention. Lines 1, 2 and 3 represent effects of intervention. Curve e is unmanageable because of the rapid dynamics; a can be stabilized but many deaths will occur before

Figure 3.19 Dynamics of unmanageable and manageable scenarios

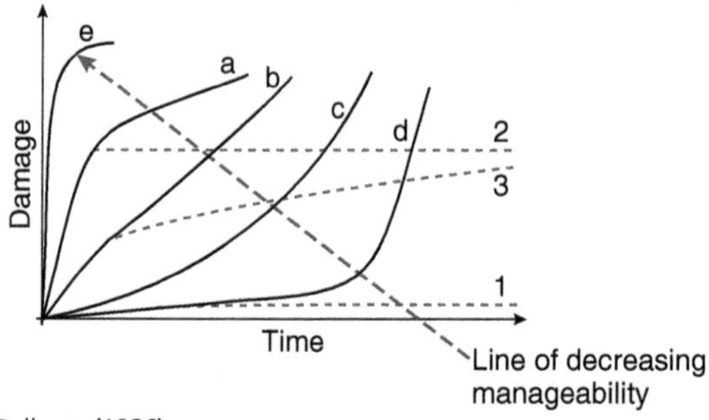

Source: Adapted from Bellamy (1986)

intervention because of the rapid development of the scenario; b cannot be fully stabilized at point of intervention, so harm goes on increasing; in c no intervention occurs; and d can be stabilized with minimal harm. For example, increasing possibilities for self-rescue (for example, by well-designed alarm and evacuation systems) reduce demands on rescue personnel.

Sheltering and evacuation

When a threat such as a toxic cloud impacts on an area and people are present, they have to escape in order to survive. Risk assessments have to take account of the effect of the threat to human behaviour such as whether self-rescue is motivated and/or impaired by the results of a toxic substance such as difficulty in seeing or breathing, or, in the case of toxic effects on the central nervous system, confusion and inability to coordinate movement.

The possibility that those caught outdoors will be able to run for cover, including the dose they would receive while running and once indoors, will influence their chance of survival. The time needed to reach shelter is a characteristic of the neighbourhood. Incident studies show that people always start to run, but, unless they are trained personnel and have a view of some wind direction indicator, the direction they will run in is almost completely random.

In addition, a significant proportion of the population may be in cars. They may also leave their cars after a certain period of time and try to run for shelter.

Include the effect of action by the emergency services

The effects of toxic releases as predicted by risk assessment models indicate far greater fatalities than generally occur in real incidents. Escape, sheltering, evacuation and rescue can be considered to account for this and therefore should be modelled in the risk assessment. In addition, risk assessment traditionally focuses only on fatalities. A quantified risk assessment (QRA) does not usually analyse the behaviour of third-party assistance in the analysis of risks. In fact, if the risk picture is highly sensitive to third-party intervention, then it could be argued that it is a very poor safety system. In that respect it would be justifiable to exclude third-party assistance from a QRA and include only self-rescue. The QRA for the Channel Tunnel, for example, did not take into account any mitigatory actions by land-based fire brigades, as occurred in the fire of 1996, for example.

Typically, risk assessment is not prepared with emergency preparedness and response in mind (conclusion of the OECD workshop on Risk Assessment and Risk Communication, Paris, 1997, OCDE/GD(97)31):

> [I]t was frequently stated that the analysis of selected scenarios from an overall risk assessment could be of assistance. The emphasis is usually placed on organisation and logistics. It is still a matter of judgement, by experts in emergency planning and risk assessment, to decide how much risk assessment methodology can and should be used in emergency planning. Another way to look at this is to formulate a risk assessment methodology that is designed for the purpose of input to emergency plans. In fact, it is probable that the needs for this purpose are so different from the needs of planning and authorising authorities that a new approach is required. However, it should not be the case that these two methods of risk assessment are inconsistent with one another.

An unmanageable scenario can be defined according to the following parameters:

- no warning (e.g. compared to a flood warning situation)
- dynamic development (harm increases over a relatively short time)
- stabilization not possible within available response times (the damage goes on increasing).

Manageable scenarios allow time for warnings to be given and for mitigating responses to be carried out, or for a situation to be controlled before it impacts on a population. These days, sheltering may be the preferred response since it has been recognized that evacuation cannot be achieved quickly or completely. People have to be warned (notification time), they cannot leave immediately (preparation time) and they have to move away from the area either on their own or via some means of transport that has been arranged (evacuation time). Evacuation is unlikely to be achieved for 100 per cent of the population (Figure 3.20).

Bellamy and Harrison (1988) used historical movement time data for evacuations of the public to model how long it would take to evacuate a population around an installation in the event of a toxic release. The movement time aspect of the model is given by a very simple formula (far simpler than the traffic simulation models used in the United States, for example),

Figure 3.20 Overlap between different stages of evacuation

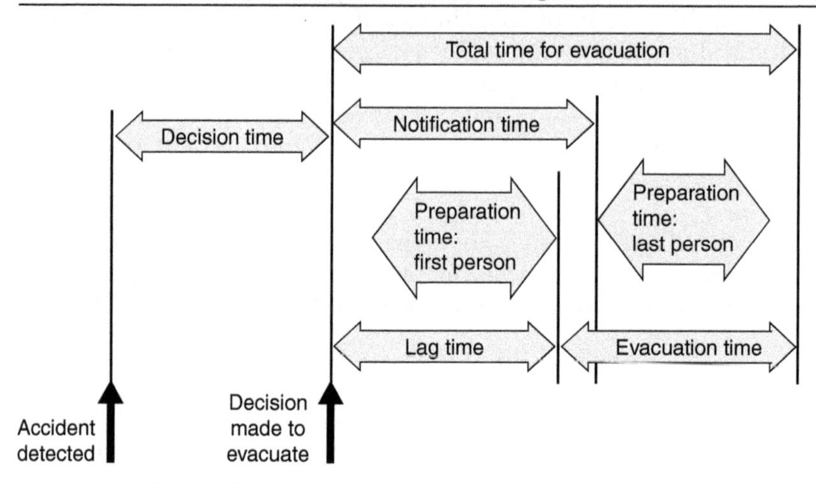

Source: Adapted from Bellamy (1988)

where evacuation rate is expressed as a function of number of people to be evacuated:

$$y = 14.12(\pm)0.5$$

where x = the numbers to be evacuated and y = the evacuation rate (numbers evacuating per hour). It has been shown to work well in validation studies, and conforms to rules of thumb used by emergency response personnel such as the police. Models such as these rely on having a large database of suitable information, but require no behavioural explanations or rules for them to work very effectively (Figure 3.21).

What is far more difficult to model is the lag time before evacuation commences. This is very dependent on local factors such as the methods of warning, the time it takes decision makers to evaluate the situation, and the decision itself.

With regard to sheltering, success is dependent upon factors such as the length of time of a release, the effectiveness of warnings, the adequacy and safety of shelters, access time, especially for transient members of the population, whether people might leave a shelter too early or get trapped there, and failure to bring to an end the need to shelter.

Many different attributes can affect people's responses to warnings and threat, such as age, desire to reunite with family members, socio-economic status, and prior knowledge and understanding of the threat. The nature and timing of the warning itself are also very significant factors. PA announcements from vehicles may cause people to go outside in order to hear them. One study in the United States concluded that warning signals were not heard because windows were closed and air conditioners and fans were being used.

Figure 3.21 Line of best fit through evacuation data points

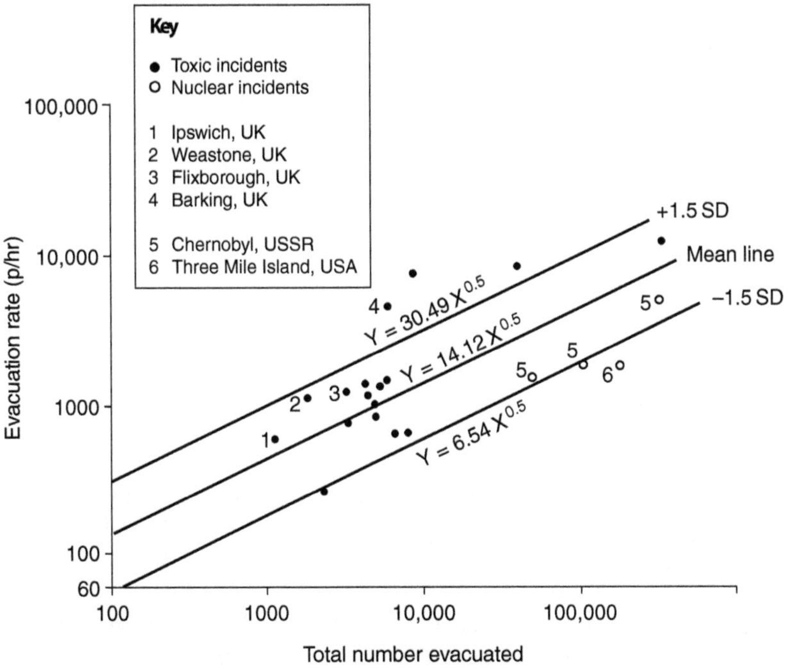

Source: Adapted from Bellamy and Harrison (1988)

In risk assessment, sheltering or evacuation may be treated as a global factor that mitigates the effect of a release without modelling this at all in any detail. However, this is not very helpful in the decision-making process, especially if the risk assessment has already been factored into the mitigation and has no parameters that can be manipulated to investigate optimum strategies.

Command support issues

Command support issues concern the timely rescue of trapped or injured persons within a period when life support can be maintained through:

- source reduction
- effect reduction
- dose reduction
- exposure reduction
- harm reduction.

Proactive assessment of the possible demands of a system (which releases a hazardous chemical, say) on these reduction requirements may be partly based on scenarios developed in QRA and partly on deterministic scenarios developed by some systematic or creative process. The reality is, however, that the range of scenarios that may be developed for evaluating the need for command support are not usually determined by a formally specified method and are highly dependent upon the skill and understanding of the analyst. Scenarios relevant for command support generated by methods like QRA may not address all the relevant issues, especially the extremely important dynamic factors and aspects related to types of injuries and other information requirements.

Ideally, evaluations should address remaining scenarios after unmanageable scenarios have been eliminated. Scenarios should be identified and developed where hazards to command support could be realized, including, for example:

- insufficient means of access and egress for rescue, including means of access for equipment;
- insufficiency in fixed resources which enable rescue to take place, or the necessary fixed facilities which enable transported resources to be used;
- impairment to timely rescue, such as might result from:
  - impairment to means of access and egress required for rescue and rescue equipment, for example due to effects of the accident, or due to insufficient space;
  - impairment or damage to any of the necessary resources required for rescue, for example problems with using protective suits against accident conditions;
  - impairment to visibility, for example due to loss of lighting or the occurrence of smoke;
- impairment to movement, for example due to overcrowding, narrow spaces or physical obstructions;
- damage to the command and coordination structure, for example through receiving insufficient or incorrect information, loss of communication or loss of necessary personnel;
- delay (in getting the alert, in arrival, in gaining access to resources, in gaining access to source of threat or victims, in getting victims to a place of safety, or stabilized;
- loss of availability of necessary personnel (not just whether they are present but also whether they are competent, and so this point also includes training, and it includes operations management personnel and not just external services);
- loss of availability of any of the necessary resources;
- impairment to communication, for example loss of radio function, or excessive noise.

Analysis should examine each stage from alert to stand-down. For example, for alert the analysis should include examining the way in which a potential emergency is identified, the equipment (detection, alarms) and which personnel are involved, and the alerting and informing of the external services.

The scenarios should be analysed in such as way as to provide information to the decision process.

QUANTIFIED RISK ASSESSMENT

All the information and tools can now be combined to result in a risk assessment. In this assessment the two quantifiable dimensions of risk – probability and consequence – are calculated and expressed in the metrics required by the decision maker and the decision-making process. For risks such as those associated with nuclear power plants, chemical factories (Ale and Whitehouse, 1984), airports (Ale and Piers, 2000) and air traffic (Ale *et al.*, 2006, 2007), there are many structured – often computer-assisted – methods available. In addition to the quantification, a qualification needs to be given. This I shall describe further in the final chapter, where we shall discuss decision making and the role of perception and qualification in that process.

4 Managing risk

In this chapter we shall talk about risk management: the organized way of keeping risks under control. We shall discuss the various methods of doing so and the models or frames that are used to come to grips with the problem of risk.

The origin of modern risk management lies in the industrial accidents after the Second World War. In 1966 a fire in a storage facility for liquified petroleum gas (LPG) in Feyzin, France, killed 18 and injured 81 people. This accident led to re-emphasis on design rules for bottom valves on pressure vessels. In the realm of physical planning, no actions on the part of the French or the European authorities seem to have resulted from that accident.

Ten years later a number of similar accidents occurred: Flixborough (1974, 28 dead), Beek (1975, 14 dead) and Los Alfaques (1978, 216 dead). These accidents showed that the Feyzin accident was not a unique, freak accident. Apparently, LPG and other flammable substances could pose a serious threat both to the workforce at a plant and to the surroundings.

Studies were commissioned into the safety of industrial risks at major petrochemical complexes and of the transport of hazardous materials (HSE, 1978, 1981; Cremer & Warner, 1981). It became apparent that a policy aimed at ensuring that no accident would ever harm the population would not be compatible with the limited space available around these sites. So, the unavoidable conclusion was that there should be a level of risk below which it is neither desirable nor economic to strive for further reduction.

As we have seen in the previous chapters, systems behave under the influence of inputs, controls, environment and other systems. This behaviour depends on how the system is put together. If we want to change the behaviour, we have to change the system, the interaction between systems or the interaction between the system and its environment. We have already seen an example in Chapter 3. In the bow tie diagram it was shown that possible types of intervention are (1) to remove a cause, or (2) prevent progression of an unwanted pathway by introducing a barrier. In this chapter we shall look at these types of intervention in more detail.

THE DEVELOPMENT OF AN ACCIDENT

There are many models for the process that constitutes an accident. (e.g. McDonald, 1972; Kjellen, 1983). An example is given in Figure 4.1.

Figure 4.1 Example of an accident model

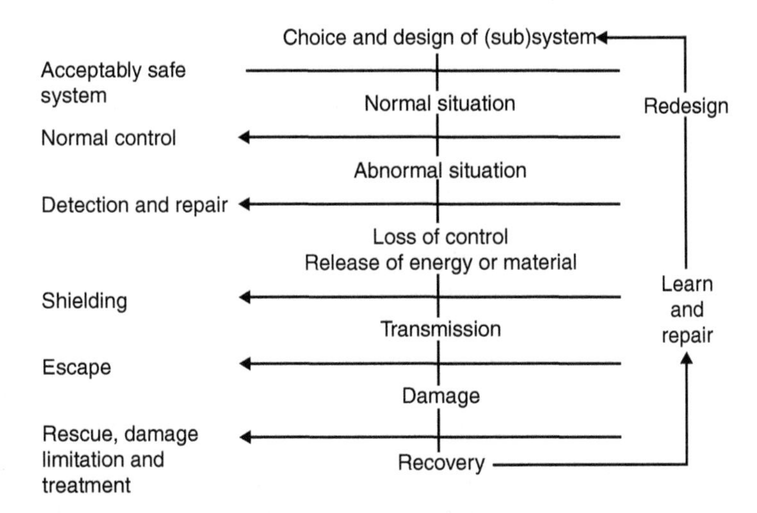

These models have the character of a description of relationships between aspects that the particular modeller thought or thinks it important to mention or make explicit. Therefore, there are many such models, all of them different. None of them is the ultimate 'right' model. What they have in common is that they try to make one, some or many actions or interventions explicit. In more recent literature (Debray *et al.*, 2004) they are called barriers.

Barriers

As I have already described, the mathematical concepts of fault and event trees prove to be difficult to handle in a managerial or policy process. Managers and policymakers seem to be much more comfortable with the concept of a 'safety function'. This safety function is identified as being separate from the process or activity at hand. Its sole purpose is to reduce or mitigate some negative or unwanted outcome of the activity. It was Haddon (1973) who introduced this barrier concept. Barriers are thought of as blocking the path between cause and consequence. Adding a barrier as a safety device is technically the same as combining a safety device with a consequence-causing event in an AND gate. In the barrier concept the causing event has to occur together with the failure of the safety device. In the fault tree concept the causing event and the failure of a safety function are not distinguished. The two events – cause and safety function failure – have to occur simultaneously for the causal chain to propagate.

The concept of barriers has been elaborated in the ARAMIS (Delvosalle *et al.*, 2004) project. The primary purpose of this project is to promote consistent decisions where scientific risk analyses give diverse and uncertain answers. These decisions pertain to the implementation of safety functions in order to avoid, prevent, control or limit the frequency or the consequences of an unwanted event.

Let me note in passing here that the ARAMIS guideline resulting from the project does not specify how the effectiveness of safety functions can be quantified other than by referring to existing methodologies and literature. This implies that the uncertainties resulting from scientific risk analyses cannot be resolved by using the ARAMIS methodology. This means that at some point a choice of preferred and adopted method or number has to be made, which puts decision makers, managers and policy developers in the same predicament as before the ARAMIS project.

The barrier approach is to a certain extent a metaphor for what the purpose of a safety measure or a safety function really is. Such metaphors are useful and in many cases unavoidable in transferring knowledge from the technical and scientific domain into the managerial or policy domain. Yet however useful these metaphors may be, they pose the danger that more and more interpretational and explanatory work has to be done to keep the two domains aligned. In this way, barrier classification and explanation becomes a science in itself and may obscure the original idea and the original purpose.

As the barrier concept is so appealing in comparison with the more rigorous fault and event tree approach, a further description is needed. Some of the effects of adding something to the chain of events are almost automatic in the rigorous approach, yet need careful consideration when using the barrier concept.

Two types of barriers

In the ARAMIS project and associated studies, various genealogies of barriers have been developed. For the purpose of the WORM project, two different barriers need to be distinguished: static and dynamic barriers.

A **static barrier** is defined as a barrier that is always present and is meant to prevent progression of the accident chain towards the consequences. Static barriers include such things as gates, edge protection, fall arresters, covers and (second) containments. A common property of static barriers is that they are physical.

A **dynamic barrier** is a barrier that needs some kind of activation. Dynamic barriers could include medical attention, emergency vents and shutdown valves.

There are several instances where a barrier is conceived as being static but in its implementation is dynamic. A safety distance may be interpreted as a static barrier; the distance is always there. But if there is no physical gate at the required distance, and the required distance is marked simply by a line on the ground or a sign, the barrier becomes dynamic: the line is a warning signal that has to be noted, identified and acted upon.

Figure 4.2 A static barrier does not need a transitional area

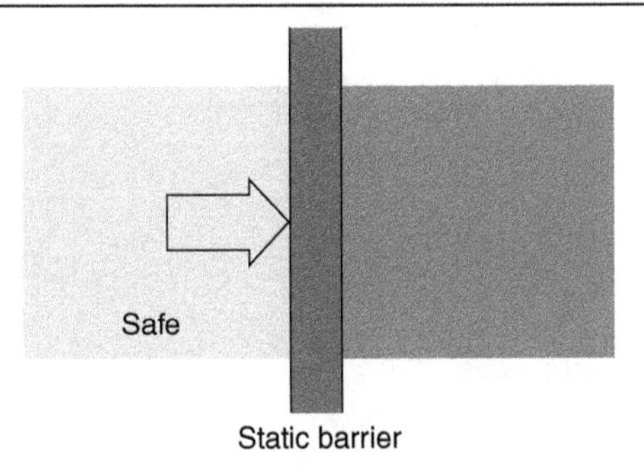

Figure 4.3 Dynamic barrier

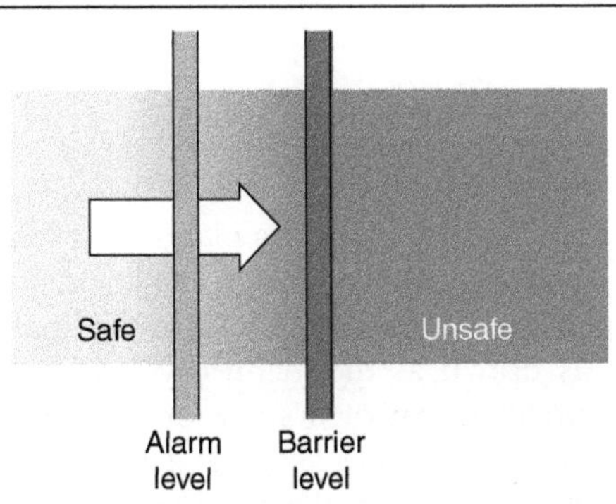

Static barriers do not need a transitional area between safe and unsafe: the edge protection can be at the edge (Figure 4.2). A dynamic barrier, by contrast, does need some transitional area between safe and unsafe. This is where the barrier is activated, when a person who is about to enter a danger zone stops before crossing the line (Figure 4.3).

From barriers to safe envelope

The idea of a safe envelope of operation has its roots in system theory. It is drawn partly from the ideas of Rasmussen (1997), who proposed a highly abstract model in which organizations try to operate within a safe 'area' defined by rules, procedures and (inherent) safe design measures, or barriers. A safe envelope of operation can be defined as a multidimensional space in which the activity or process (in the case of a machine, the feeding and operation of the machine) takes place without damage occurring. Figure 4.4 shows the safe envelope concept in diagrammatic form.

If the boundary of the safe envelope is breached, damage is unavoidable, though its extent may still be subject to influence by further mitigating measures. In terms of the earlier bow tie representations, the loss of control may be just inside the safe envelope boundary (SEB), if there are still risk control measures that can prevent injury or damage; or it may be on the boundary

Figure 4.4 Safe zone–danger zone personal protection: machines

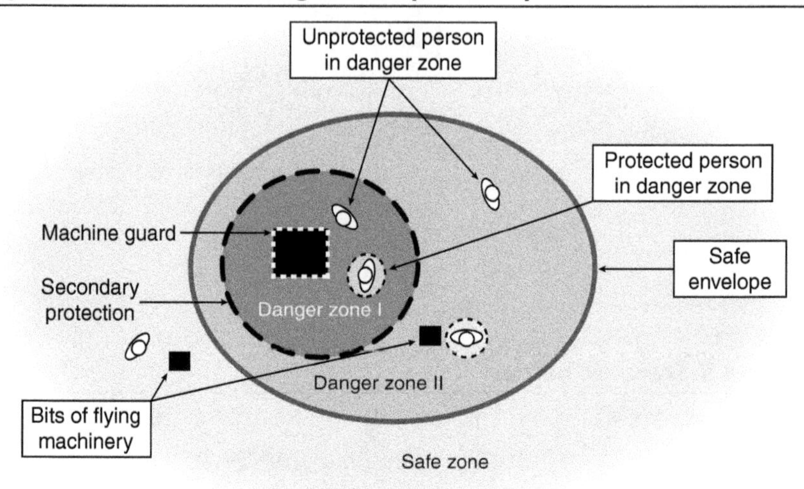

itself, if there are none; or there may be a margin in which the system is out of control, but chance or outside intervention can bring it back before it crosses the SEB. This is indicated by the dotted line for the boundary of controllability. An inner line is drawn as the 'defined operational boundary' (DOB), within which the organization or system, or the person in control of the online risk management, wishes to keep the activity, so that there is a safety margin to increase the level of actual safety of operation.

For a ladder, the safe envelope is defined among other ways by ensuring that the centre of gravity does not extend beyond the support point (Figure 4.5).

Boundaries are often not exactly known and very often are not fully visible to those trying to operate within them (or to those assessing and regulating the operations). Risk control measures are defined, in the first

Figure 4.5 Safe envelope for a ladder

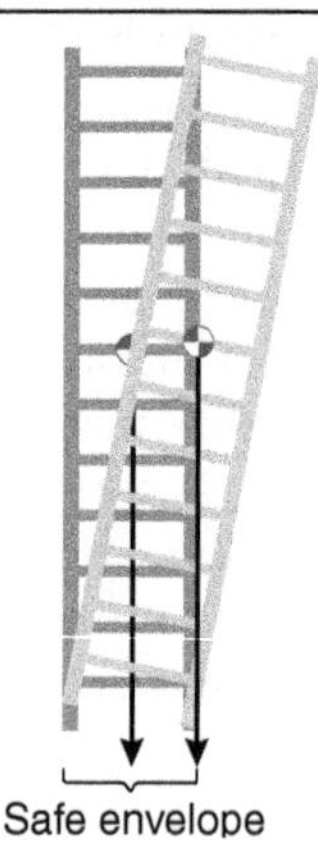

instance, as measures which ensure that the operation stays within the SEB, and preferably within the DOB, by **detecting, diagnosing** and **acting** on information that the boundaries are being approached and will be crossed if no action is taken. It is important to define the boundaries as physically and as deterministically as possible, so that they can be modelled unequivocally. These are risk control measures in the classic sense of barriers. In many systems the SEB and/or the DOB may be marked by physical barriers such as fences, walls, dykes or personal protective equipment, or by a symbolic warning in the form of a white line. In the case of a machine the danger zone – the moving parts of the machine, or a crane lifting and swinging its load through a given space – is surrounded by a safe zone for the people feeding the machine or working on the construction site.

The boundaries for different activities may be tight or loose. For machines, maintenance may be an activity with narrow margins of safety. Boundaries to danger zones may also be shifted deliberately during an activity as part of the risk control. In feeding a press, an interlock guard prevents movement of a press while the guard is open, so that the operator can reach into what is otherwise the danger zone and place a component to be pressed and then close the guard, reducing the safe zone, while the machine is in operation. Hence, a control sequence is also necessary to detect that the boundary is shifting over time.

We shall now explore how we could define safe envelopes for a number of examples.

Safe envelope for a ladder

One of the modes of failure for people using a ladder is that the ladder falls over. A ladder falls over when the centre of gravity of the ladder, including a person or persons standing on it, moves beyond the base of the ladder. So – ignoring the friction between ladder and the wall against which it is set up – the safe envelope can be defined as indicated in Figure 4.5. This also gives us a clue as to what we can do to make the situation safer; that is, we can extend the safe envelope by attaching a side bar to the base of the ladder, effectively making the base wider.

We could also attach the ladder to the support structure at the top of the ladder. This would lead to a whole new definition of the safe envelope. But somebody must make the attachment first.

Safe envelope for a bridge

Suppose you have to define the safe envelope for a truck to pass through an eighteenth-century drawbridge (Figure 4.6). It is important that the truck not hit the structure. So, the height of the vehicle should be below the height of the bridge and the width of the vehicle should be no wider than the opening through which the truck has to pass. There are other parameters to

Figure 4.6 Safe envelope for a bridge

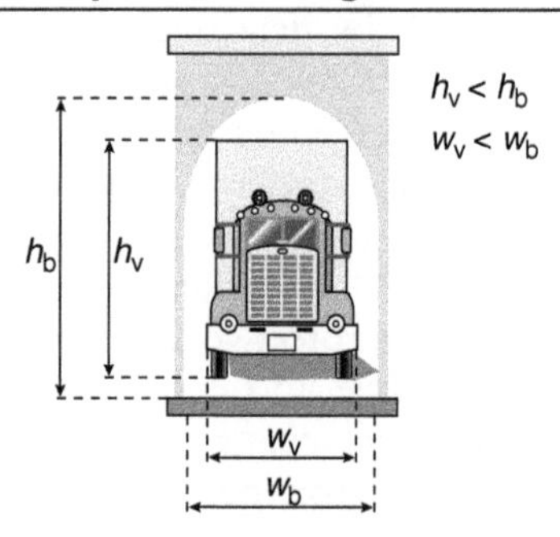

consider, though, and usually they are given in the traffic signs in front of the bridge. Maybe the reader would like to consider what these might be.

Safe envelope for an edge

When we approach an edge such as a river bank in a car, the safe envelope is determined among other ways by our capability to brake and come to a standstill before we drive into the water (Figure 4.7). The safe envelope can be derived as follows. Let v_0 be the initial speed of the vehicle and d_{max} the maximum deceleration of the car. Let s be the remaining distance to the edge. Then, the time to reach a standstill is given by

$$t = v_0/d_{max}$$

and the distance then equals

$$s = \int v dt = \int v_0 - d_{max} t dt = v_0 t - ½ d_{max} t^2 = \frac{V_0^2}{d_{max}}$$

which gives us the minimal necessary deceleration. The car's maximum deceleration should be larger than that:

$$d_{max} \geq \frac{V_0}{2s}$$

Figure 4.7 Safe envelope for an edge

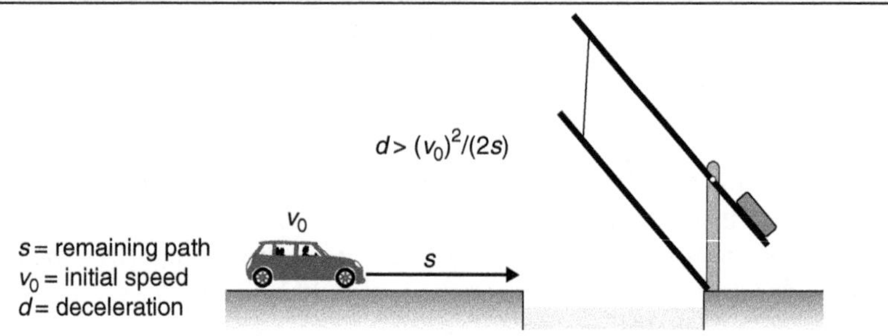

As we can see from these examples, the definition of the safe envelope in physically defined parameters can be simple, such as 'keep a certain fixed distance', or complicated: a set of parameters and equations that define the boundary of safe operation. Transgressing one of these parameters or boundaries leads to a situation in which an unwanted or undesired effect can, or even will, happen, depending on the number of boundaries that have to be crossed before an accident is no longer avoidable.

In terms of the barrier description, the transgression of a boundary is equivalent to the breaking of a barrier. In order to keep a system safe, these barriers have to be set – technically – and known to the user of the system, or to the people within the system. To have these barriers in place and to keep them working is what constitutes safety management.

MANAGEMENT OF RISK

There are a whole series of models that are meant to capture the necessary components of a risk management system and the way it keeps a system safe. These models are different from models in physics in that they are often merely a description. Not much of management theory in general and safety management in particular has been tested in controlled experiments. The model described here has been developed over many years (Bellamy *et al.*, 1993; Papazoglou *et al.*, 2003; Delvosalle, 2004; Ale, 2006).

Barrier tasks

In this model we distinguish four tasks that cover all activities that need to be done to have a barrier and keep it functioning. These tasks are:

- provide
- use
- maintain
- monitor.

Each of the four tasks has to be fulfilled to keep the barriers functional (Figure 4.8).

Figure 4.8 The four barrier tasks

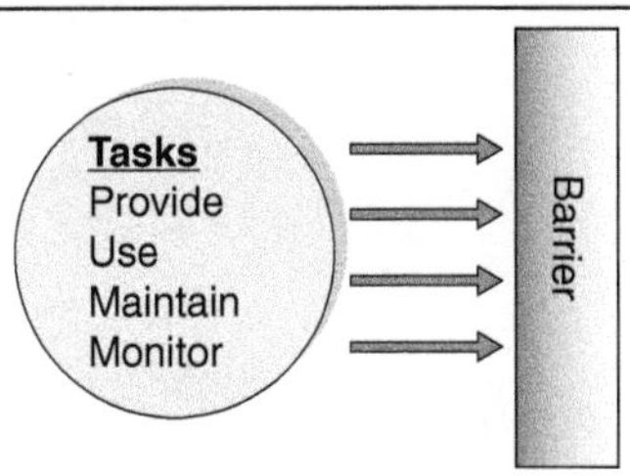

Provide

'Provide' stands for making available, constructing, installing and all other synonyms and descriptions that have the connotation of having a barrier present.

Use

Once provided, the barrier has to be used. This task obviously relates to operators and other human beings, who have to use the extant safety provisions. It also pertains to machines and computer programs, where safety provisions could be switched off or circumvented, which should not be done if the barrier is to work.

Maintain

'Maintain' covers all activities necessary to keep a barrier in good and appropriate working order. These activities may include installing updates to technical developments, making necessary adaptations to changes in working conditions and making necessary changes to the operation of systems and operating environments.

Monitor

'Monitor' covers all activities associated with observing that the barrier is functioning correctly, including inspections, audits and tests.

Deliveries

The deliveries are the components of working safely. They have to be 'delivered' to the barrier by the barrier tasks described above. For instance, for a barrier to work correctly there have to be procedures, and these have to be provided, used, maintained and monitored. Eight components of working safely can be distinguished:

- procedures
- equipment
- ergonomics
- availability
- competence
- communication
- commitment
- conflict resolution.

I shall now describe these in turn.

Procedures and plans

Rules and procedures are specific performance criteria that specify in detail, often in written form, a formalized 'normative' behaviour or method for carrying out an activity. 'Plans' refers to the explicit planning of activities in time, either how frequently tasks should be done, or when and by whom they will be done within a particular period. They include the maintenance regime, maintenance scheduling and inspection activities.

Equipment, spares and tools

This component refers to the equipment and spares that are used and installed during maintenance. The spares and tools must be correct ones and they must be available where and when they are needed.

Ergonomics

'Ergonomics' refers to the quality of the hardware itself and, in many advanced systems, the quality of the software that controls it, and the ergonomics of the interfaces with them that are used or operated by operations, inspection or maintenance. It includes both the appropriateness of the interface for the activity and the user-friendliness of the facilities that are used to carry out the activities.

Availability

'Availability' refers to allocating the necessary time (or numbers) of competent people to the tasks that have to be carried out. This factor emphasizes time-criticality; that is, people must be available at the moment (or within the time frame) when the tasks are to be carried out. This delivery system is the responsibility of personnel planning. A critical aspect is planning for peak demands, particularly in emergency situations or at other times when deviations from normal or planned operations occur.

Competence

'Competence' refers to knowledge, skills and abilities in the form of first-line and/or back-up personnel who have been selected and trained for the safe execution of the critical tasks and activities in the organization. This system covers the selection and training function of the company, which must deliver sufficient competent staff for overall personnel planning. Competence should be seen as not only cognitive but also physiological; that is, it includes factors such as health and fatigue.

Communication and coordination

Communication occurs implicitly or explicitly within any task activity when it involves more than one person. Proper communication ensures that the tasks are coordinated and everyone knows who is doing what. Communication and coordination are particularly critical at shift changeovers, when an aircraft is handed over from operations to maintenance and back again, when new or temporary staff are involved, or when transferring work between departments in a complex organization. This delivery system is diffused throughout the organization and includes meetings, permit to work systems, protocols and plans, logbooks, etc.

Commitment

'Commitment' refers to the prioritization, incentives and motivation that personnel need in order to carry out their tasks and activities with suitable care and alertness, and according to the appropriate safety criteria and procedures specified by the organization. This delivery system deals with the incentives of individuals carrying out the primary business activities not to choose other criteria above safety, such as ease of working, time saving, social approval, etc. The delivery system is often diffuse within companies, but includes many of the activities of supervision, social control, staff appraisal and incentive schemes. These apply all through the organization from top management to the flight deck, control room and shop floor.

Conflict resolution

'Conflict resolution' includes the mechanisms (such as supervision, monitoring, procedures, learning, group discussion) by which potential and actual conflicts between safety and other criteria in the allocation and use of personnel, hardware and other resources are recognized and avoided, or resolved if they occur. This covers the organizational mechanisms for resolving conflicts across tasks and between different people at operational level and at management level.

The Russian Doll problem

The problem with analysing management systems is similar to the problem with analysing any system: one can always further analyse the subsystems into other subsystems.

For instance, one has to use the right personnel for the barrier. To make this happen, one could write a procedure on how to get the right personnel. And one could maintain that procedure ...

This mechanism is present not only in the analysis of management systems but also in safety management itself. This is one of the factors that cause legislation, codes and regulations to resort to ever-increasing complexity. If something is regulated, it becomes necessary to regulate how regulations have to be maintained, and these new regulations have to be discussed with stakeholders.

For reasons of simplicity, one could dismiss the deeper layers of this construct and look only at the top. However, as we have already seen in some examples and will see in others, often the onset of an accident lies in the deeper layers. This means that either we have a mechanism in this top layer that catches all deviations that occur in the deeper layers, or we have to take these deeper layers into account.

Material barriers

Material barriers are physical entities such as shields, barriers and devices that prevent a fault, an error or a deviation from further developing into an incident or accident.

Shielding

As we have seen earlier, probably the most visible of safety functions or safety provisions are physical barriers. They are there to prevent contact between a human and a hazard. In the case of machines, these are the machine guards; in the case of chemicals, these are the closed containment; in the case of a highway, the central crash barrier prevents contract between the two streams of traffic. A safety net under a scaffold prevents falling objects from hitting people.

Correct operation

There are less visible material barriers. These are all the provisions made to make a system work normally: to make a machine produce the right products, to make a robot make the right moves, to make a chemical factory produce the final chemical. All these provisions – even if they are primarily meant to make the system reliable and profitable – also serve the goal of safety. We shall see later how important it is to keep this in mind when we come to discuss management of change, cost–benefit analysis and risk control.

Immaterial barriers

Immaterial barriers are aimed at making people do the right thing, which in the case of safety means making them do the correct thing. They include rules and regulations. An example is the rule about keeping enough distance between you and the car in front. There is nothing physical that prevents you from tailgating the car in front of you, but doing so is recognized to be highly dangerous. The safe operating envelope is defined by the maximum deceleration of the car in front of you, the speed of your reaction when the brake lights of the car in front of you light up and the deceleration of your own car. It is generally not possible to know exactly how large the distance between you and the car in front of you should be. Therefore, a rule is set, such as 2 seconds or 100 metres, or, as in Belgium, 'three chevrons'. The only thing that keeps you from crossing this barrier is you.

With this type of barrier, motivation and competence are much more important components than equipment or even conflict resolution.

Barrier failures

The most likely type of failure concerning material barriers is that the barrier is not provided in the first place. Not using the barrier when it is provided involves physically removing the barrier or taking some action to circumvent it, such as climbing over a fence into a vessel after removing the lid or cover. The reasons for circumventing a barrier are likely to involve motivation and, to a certain extent, competence – the latter associated with the understanding that the barrier was there to protect people from harm.

For immaterial barriers the situation is more complicated because these barriers involve a detect, interpret, act cycle (Figures 4.9 and 4.10). This cycle may be embedded in a system such as a process computer or an automatic lockout. In such a case the safety function resides in a 'black box',

Figure 4.9 Dynamic barrier control cycle

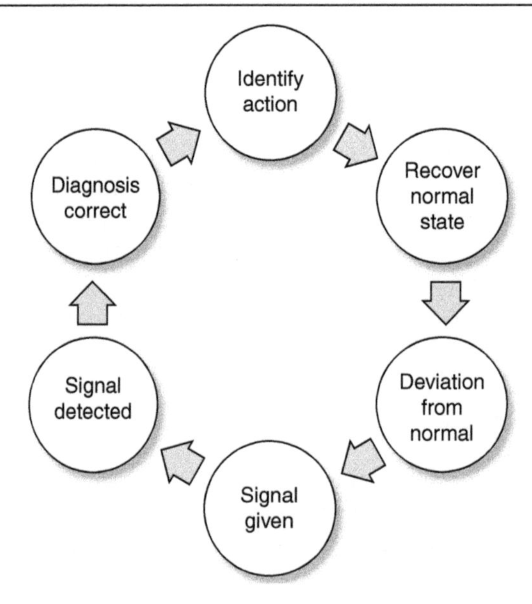

Figure 4.10 Event tree for a control cycle

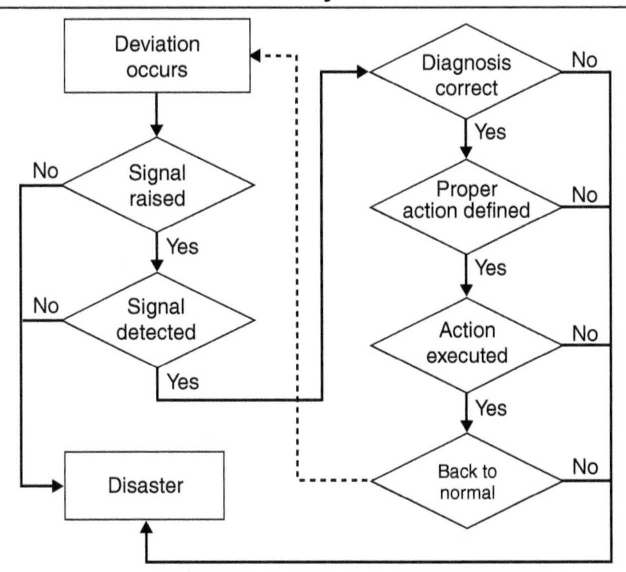

which is usually interpreted by an analyst as constituting one single barrier. The analyst will most likely take the inner workings of this box as a second-order problem, if he or she considers it at all.

When the control cycle involves human action, such as hearing an alarm or seeing a line on the floor, the separate components of the cycle tend to appear in the analysis separately and explicitly. This also has an effect on the task and delivery failures that are likely to be identified by the analyst. With the 'system'-controlled dynamic barriers it is again more likely that a provide failure will be noted, while in human-controlled barriers the tasks identified are more likely to be use, maintain and monitor.

Control loops and learning cycles

We have seen that the management of risk is all about detecting deviations from the norm or the wanted and acting accordingly. This does not merely apply to the immediate system of Man, Technology and Environment such as a hazardous process or a dangerous work activity. These systems are all part of the larger system constituted by a company, which in its turn also is only part of the larger fabric of society. Therefore, control loops should be present not only at the level of the individual worker, but also at the level of the company and of society as a whole. If these control loops are broken, the road to an accident is opened. IRISK (Oh *et al.*, 1998) identified four layers of control or learning loops, namely the rules, the regulation, the laws and the climate in which the system operates. IRISK laid particular emphasis on organizational and societal learning. Thus, in IRISK it is regarded as important that the learning cycles are also implemented within the culture of the system as a whole.

Figure 4.11 The four layers in the IRISK risk control system

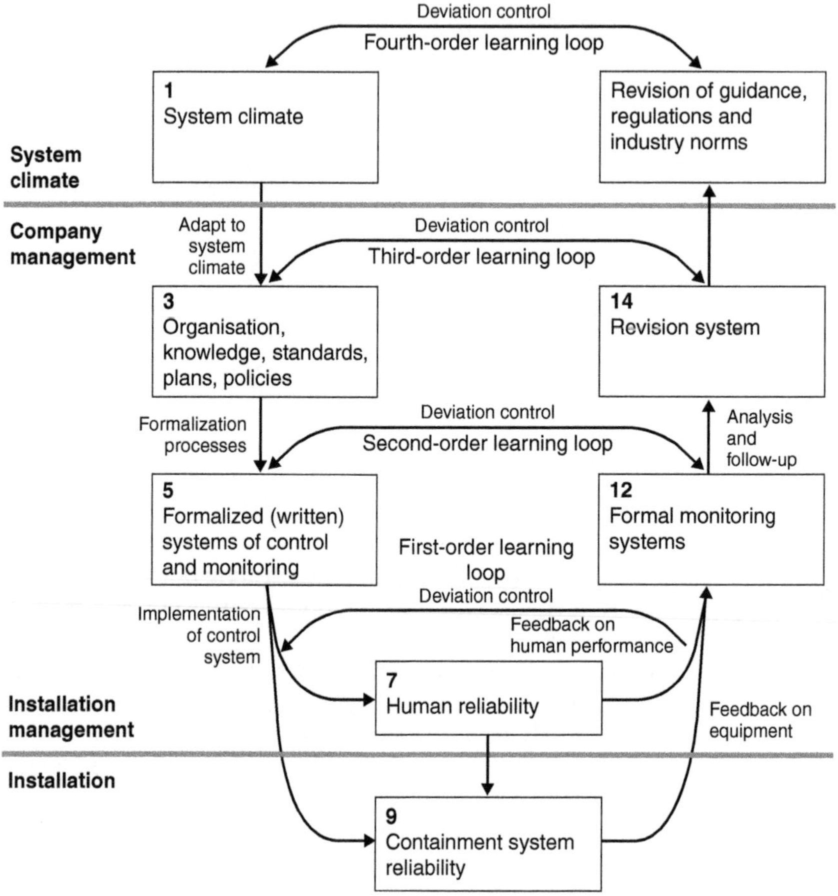

Deviation control
Fourth-order learning loop
1
System climate
Revision of guidance, regulations and industry norms
System climate
Company management
Adapt to system climate
Deviation control
Third-order learning loop
3
Organisation, knowledge, standards, plans, policies
14
Revision system
Formalization processes
Deviation control
Second-order learning loop
Analysis and follow-up
5
Formalized (written) systems of control and monitoring
First-order learning loop
12
Formal monitoring systems
Deviation control
Implementation of control system
Feedback on human performance
7
Human reliability
Installation management
Feedback on equipment
Installation
9
Containment system reliability

Source: Adapted from Bellamy *et al.* (1993)

Other, similar descriptions have focused more on the judgemental processes that are involved in setting up these standards, rules and regulations, such as the model proposed by Svedung and Rasmussen (2002). They distinguish five layers of control, but close inspection of Figures 4.11 and 4.12 reveals that this is only an apparent difference. IRISK regards laws and regulations as one layer, Svedung and Rasmussen regard it as two, however. The emphasis in the latter model is on judgement. Svedung and Rasmussen emphasize that the outcome of an analysis or an inspection or the influences of society does not and should not result in predefined changes to rules and regulations, or changes in denial or approval of a permit to work. This is opposed to models in which the results of an analysis are compared with a predefined standard or criterion and in which the action taken is directly determined by this comparison.

Figure 4.12 Control as a process involving judgement

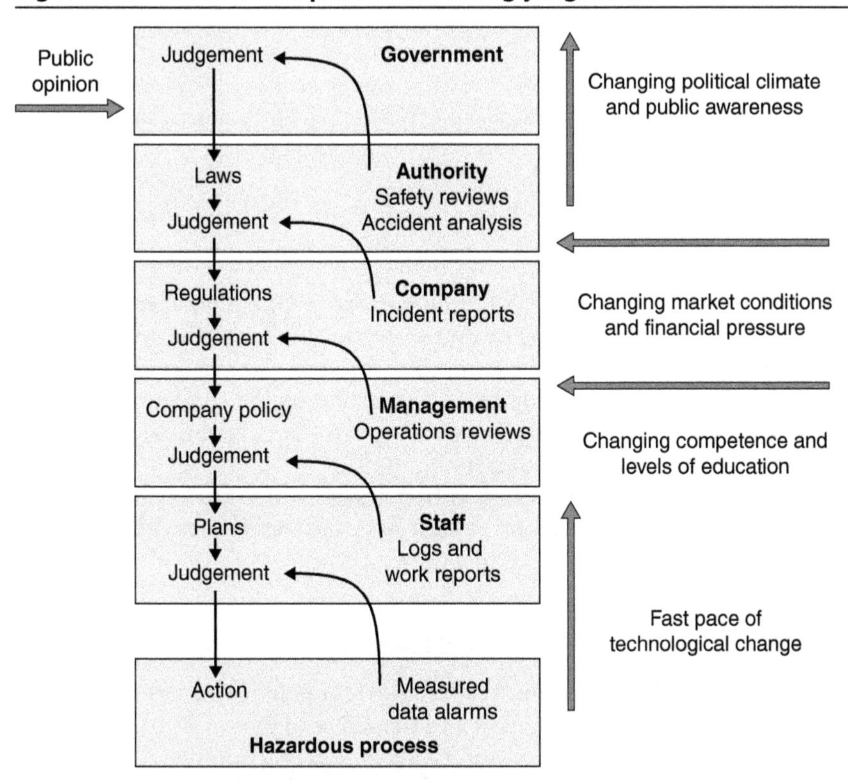

Source: Adapted from Svedung and Rasmussen (2002)

Event tree instead of barrier

Control cycles and their fate are much more easily described in the form of event trees. For the control cycle of Figure 4.9 the event tree could look as given in Figure 4.10. The advantage of this approach is that it is straightforward to add more elements and thus give more detail to the working of the control cycle, whereas these additional elements are buried in the 'barrier' in the barrier representation. For quantification, the event tree representation is used, as we shall see later.

Similarly, adding a barrier is identical to adding an AND gate in a fault tree (Figure 4.13).

T.O.C.: UNSAFE ACTS AND UNSAFE DESIGN

As we have seen, accidents happen because barriers are broken or because all the conditions for an accident are in place. It is tempting to consider the act immediately prior to the accident as the cause. If it is the act of a human,

Figure 4.13 Barrier and AND gate equivalence

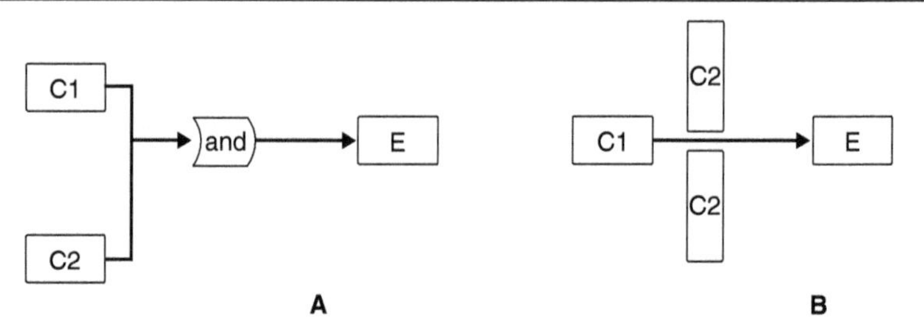

we blame that individual and call it human error. Often, however, the error was made because the making of that error was built into the system in the design stage. In order to understand this, we need to have another look at the life cycle of a system (Figure 4.14). Errors, faults, incidents and accidents usually occur during the operation phase, but the underlying causes are often found in the earlier phases.

In many cases the basic cause is bad design. There are various ways in which a design can be faulty, or at least have unsafe features. These can be paired with the list of deliveries mentioned earlier.

Incorrect choice of technology

Before design and construction really begin, some fundamental choices are made that influence the economics and the safety of the whole of the further life of the system. For instance, for an alkylation reaction one may use sulphur trioxide (SO_3) or hydrogen fluoride (HF). Both chemicals are dangerous. They are both extremely toxic and cause chemical burns when they come into contact with the skin. SO3 can form a toxic cloud at short distances from a storage tank, which makes this chemical a problem to use near populated areas. However, HF when released to the atmosphere will travel larger distances at dangerous concentration levels. This is why companies such as DuPont now favour SO3-based alkylation. Once the HF route is chosen, the problem of the risks for large areas around the plant will not go away, and choosing this technology will therefore have a continued effect on the building, operating and decommissioning of a plant.

It is probably obvious that the design cannot easily be changed if a design error is found in the operating phase – and even in the building phase it is already difficult to make more than small changes in the design. Therefore, it is important that risk evaluations be done during the design phase, so as to bring potential hazards and risks to light in order that choices can be made in time.

Figure 4.14 Life cycle of a system

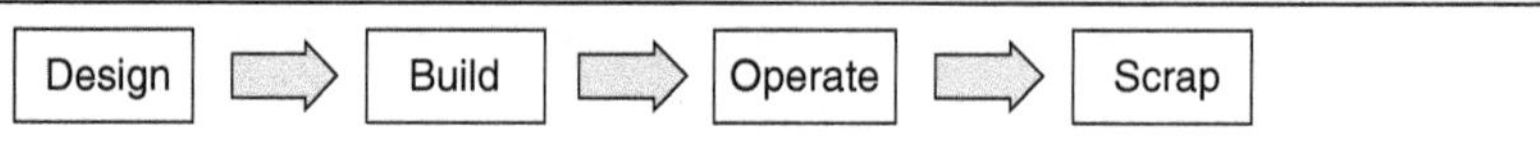

Faulty construction

Another source of long-term problems can be that the system has not been built according to design.

In June 2003 the new terminal 2E was opened at Charles de Gaulle airport in Paris. On 23 May 2004 this terminal partly collapsed, killing four people. The cause of the accident was that additional holes were made in the side of the terminal to allow for more avio-bridges. This weakened the construction. During the 11 months between the terminal's construction and its collapse, cleaners had found pieces of concrete on the floor, but nobody noticed that the load-bearing beams were slowly sagging, until the point of collapse was reached (Figure 4.15).

As construction according to design is important for the correct functioning of a system and for its continuing safety, it is important that the construction process be monitored, and continually checked to ensure that the construction is in accordance with the design.

Faulty ergonomics

Many errors, incidents and accidents are the result of sub-optimal ergonomics. Sometimes the way the system presents itself to the user is ambiguous or confusing; sometimes it can also be the case that the working of controls is counter-intuitive or abnormal. But an abundance of signals, dials and controls can also hinder effective action in the event of an emergency. It can also

Figure 4.15 Partly collapsed passenger terminal at Charles de Gaulle Airport, Paris

Source: AP/reporters

be the source of an emergency. Particularly when operators are under time pressure or have to perform multiple tasks simultaneously, the probability of error increases dramatically.

Human error

Confusing information and time pressure both increase the probability of human error because humans seldom think. They mostly act on the basis of existing, automated skills (Reason, 1997). This is why people sometimes end up at their place of work when they are supposed to be on holiday. We turn right every morning, and once this turn is taken, we travel the rest of the way on autopilot. Reason distinguishes three levels of operation of the human brain: the automatic or skill level already mentioned, the rule level and the knowledge level.

On the rule level we try to find our way out of a mess created by our skill-based behaviour by finding an applicable rule now that our skills are no longer sufficient. This is the typical behaviour also associated with Sat Nav route planners. Once you are on a smaller road, the machine will insist that you either turn back, so that you are back on the route where the skill applies, or go to the nearest main highway, where the pathfinder can make better calculations.

Only if no rule is to be found do we resort to using knowledge; that is, use our brain for things it is good at. We start to consciously examine the situation; we try a few possible alternatives (mentally or in reality) and work out what to do next. However, we do this only when we really have to, and when we are distracted in this process we return to a lower level of functioning.

Safety and culture

The way an organization deals with risks is also explained by the general attitude of the people working in the organization and its leadership. Hudson distinguishes five levels of civilization in the attitude of people and their leadership. The levels are closely related to the amount of trust the organization places in the people working in it and the amount of information that is generated, evaluated and shared between all the people and all levels in the organization. This determines the way things are done and is called the organization's culture (Figure 4.16).

The pathological level

The lowest level is the pathological level. On this level, people do not really care as long as they or the organization are not caught in a criminal procedure – in other words, as long as the lawyers are happy.

Figure 4.16 Levels of civilization of an organisation and its people

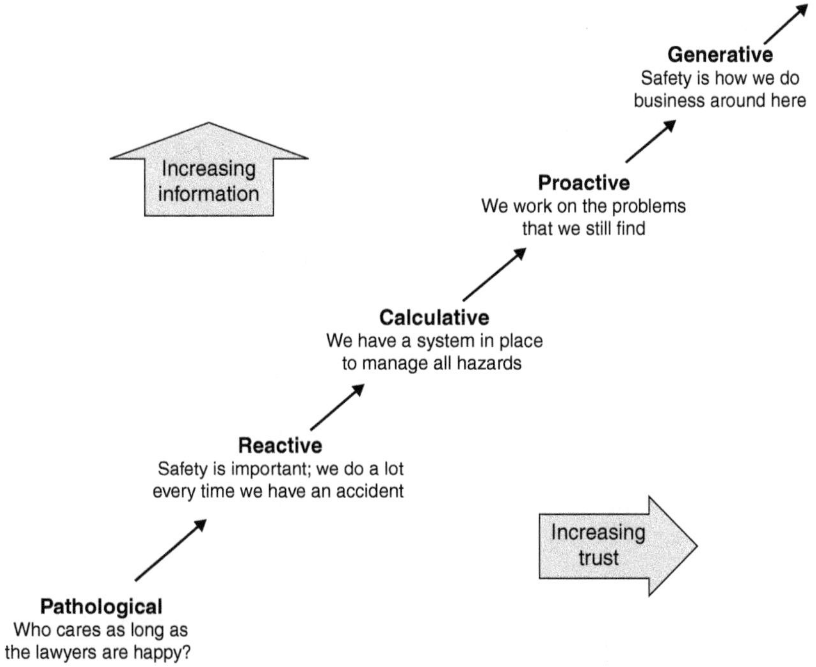

Source: Adapted from Hudson (2001)

The Reactive level

A more civilized approach is to at least do something when an accident has happened. This is the behaviour of an organization that fundamentally thinks that nothing really can happen and that risk analysts are mainly pessimists. Such an organization regards safety as important in the sense that it takes measures to prevent past accidents from reoccurring. Often these measures are also short-lived because they will be in place only so long as the organization and the people in it remember the accident and what the measures are for, which is usually not very long.

The Calculative level

On the calculative level, money is the driving force of people's actions. People realize that there are risks and that risks cost them money. In this stage of development, an organization assesses its risks, does a cost–benefit analysis and reduces risk in so far it is cost-effective for it to do so. That accidents happen from time to time is taken as an unfortunate side effect of the business and the price to be paid for progress.

The proactive level

On the proactive level an organization actively looks for potential problems and fixes them when they are found. Cost is not so much an issue any more because the organization has taken to heart the lesson that if you think safety is expensive, then try an accident.

The generative level

On the highest level, people in the organization communicate freely about problems and trust each other within and across every layer of the organization. Things are done safely or not at all. The organization does not need wake-up calls in the form of accidents to maintain the safety barriers and devices it has in place and to obey procedures. In fact, it rarely needs to refer to these procedures, because they are an embedded part of how things are done.

This level is very difficult to achieve. For most if not all companies the proactive level is the maximum obtained so far. A few companies proclaim openly that they are striving to attain the generative level.

The more convenient way to operate is also a less safe way

The convenient way is the lowest level of civilization: as long as you are not caught. Convenience is probably one of the most frequently occurring reasons why people cross or bypass safety barriers, violate rules and regulations, and in general take risks. Therefore, in each and every design the designer should ask him- or herself whether there is a shorter, faster way to do the same thing or whether more product can be squeezed out of the installation if a few safeguards are taken away. If it is done differently. If another tool is used. If do-it-yourself modifications can be made, unnoticed by any supervisor. Any designer should realize that time is the one commodity that everyone has only a limited supply of, and each and every one of us tries to use that time as efficiently as possible. This behaviour also serves as an illustration of how we perceive and deal with risks, which is the subject of the final chapter. First we shall look at the way we can use accident information to gain an insight into the structure of the risk we are dealing with.

A bank as a system

Now let us look at a completely different system: a merchant bank.

Any bank has the problem of what to do with the short-term cash it holds. This is money that cannot be invested for the long term, such as in real estate or government bonds. Usually the reason is that the bank needs to have a minimum level of cash available to pay to clients who need it, and that money came in for instance because the term of an investment ended prematurely. Banks try to maximize their profit, and often a bank has a department that handles its short-term investments. This department's task is to maximize the profit on the short-term cash. It does so by buying and selling stocks and bonds on the stock exchange. A secure investment is an investment in a company with a steady profit. These shares do not go up or down in value much, and purchasers take a share of the profit, called a dividend, as the return on their investment. The less secure stocks are in companies whose profits vary more widely. Also, investments in products such as oil or the notorious 'pork bellies' have a variable profit margin. Profits made by trading in these stocks or commodities can be high if they are bought and sold at the right time. Unfortunately, the risks involved are also high. The risk in this case is the probability of a loss. Depending on the stock or commodity involved, the probability can be high, the amount of the loss can be high, or both.

It is obvious that the expectation of the profit should be positive, and a bank, being a bank, can usually work with the expectation – that is, probability multiplied by consequences – as a measure of risk. However, if the bank has a catastrophic loss, it could become insolvent. This means that even if the bank sold everything it owned, the proceeds would not be enough to cover its debts.

Therefore, traders cannot be allowed to operate on their own. The safety system is the 'back office', which monitors the traders (Figure 4.17).

There is a good reason for having an independent safety barrier. A trader's income relies on his or her performance. So, the trader can be

Figure 4.17 The safety system of a bank, and the barrier involved

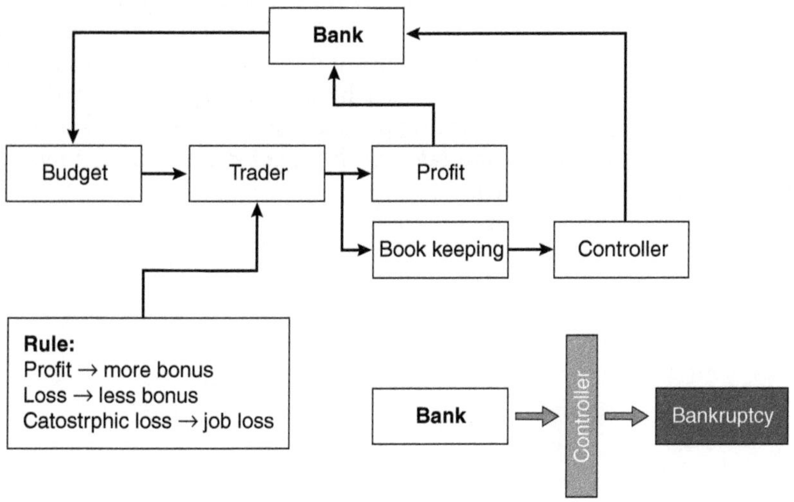

tempted (1) to take severe risks to gain high profits; (2) to try to compensate for a loss with an investment that may cover this loss (but is even more risky); and (3) failing 1 and 2, to cover his or her losses by creative book-keeping (for example, by postponing the reporting of losses in the expectation that they will be covered the next day). In short, a trader may turn into a gambling addict, risking the bank's assets.

Now suppose that the controlling function is performed by the trader him- or herself. That would make the barrier ineffective and the bank would be exposed to the risk of bankruptcy without any protection. This happened to Barings Bank, the oldest merchant bank in London, which had existed since 1762, in 1995. The controller was ill, and thus Nick Leeson, the trader, was his own controller. At first Leeson was very successful, but, as always, Murphy's law struck again and he had some very bad losses. In the end he lost $6 billion. The bank became insolvent and was bought by another bank for $1. A number of issues described earlier can be seen here again: safety functions must be provided, maintained and monitored, and they must be used. The personnel must be competent and it should be clear what has priority when there is a conflict of goals. The personnel should be committed to safety. This illustrates that theories and practices developed in one industry, in this case the chemical industry, can be applied in another, and if they had been applied, the company might still be in existence.

Bank collapse: the larger picture

The collapse of a bank may appear to be a purely financial problem, to be dealt with by financial people. However, when large mortgage providers in the United States collapsed in the early years of this century there was more at stake than just money. Many people could not pay the mortgage on their home any more and became homeless. Millions of people changed from being middle-class and well-to-do to being poor and of a lower class. So, the banks involved were dealing not only with the risk of their own bankruptcy, but also with risks to the welfare and livelihood of their customers. In fact, in the end they appeared to have been gambling with the welfare of the world at large, as their crisis provoked a worldwide financial crisis at a time where the effects of oil shortages of shortages in precious commodities were starting to be felt as well.

5 Risky information: the difference between data and knowledge

No model can be used without data. These data can only be derived from the past. In the model, parameters and variables may be changed for predictive purposes. We may in some cases be sure how certain numbers will change in the future, but we could instead be uncertain and have to work with estimates. We have to ask ourselves to what extent historical data are relevant for the present and for the future, and what to do if they are not fully relevant.

In the following we are going to look at data and statistics and see what we can do with them. We shall see that it is not sufficient to know the number of mishaps or accidents; we also need to know the exposure. After that we shall discuss what to do when there are no data available or if data cannot be found.

OCCUPATIONAL ACCIDENTS

Occupational accidents have been the subject of government policies in almost all countries in the world. And since spending government money on policies always has to be justified, much information has been collected over the years. One would expect these data to be useful, but as we shall see, even with such a common problem, using the collected data for modelling and prediction exercises has to be done with caution.

Statistics about occupational risks

The people of the Netherlands spend 7 million person-years per year at work. Each year some 167,000 persons are injured to the extent that they need at least a day's absence from work. The number of deaths is 84 per year on average.

The previous paragraph gives us several elements that are needed to say anything meaningful about the risks of activities, hazards and dangers. It has the number of occurrences of a certain event, in this case the number of injuries and deaths. It also has the exposure, in this case the number of person-hours spent on the job. This means that we can evaluate the risk of death and the risk of an injury involving at least one day off work. In the Netherlands the exposure data are usually missing.

Table 5.1 Occupational accidents in the Netherlands, 2004

2003	Total	Agriculture, Fishery	Industry, mining	Building	Transport, storage, communication	Other	Unknown
CBS	83	12	21	22	8	17	3
RIVM	104	13	23	23	12	24	9
Number of employees	7,869,000	219,000	1,046,000	465,000	467,000	5,415,000	257,000
Accident rate (CBS)	10.5	54.8	20.1	47.3	17.1	3.1	11.7
Accident rate (RIVM)	13.2	59.8	22.0	49.5	25.7	4.4	35.0

Sources: Centraal Bureau voor de statistiek, cbs statline (www.cbs.nl) and RIVM Nationaal Kompas volksgezondheid (www.rivm.nl/home/kompas)

Table 5.2 Occupational accidents in Belgium, 2004

Sector	Number of employees	Hours' exposure	Number of accidents	OBO	DO	F1	F2
Building	72,982	108,486,713	4,813	516	10	44.36	137.02
Wood	12,380	19,074,991	878	39	0	46.03	0.00
Metal	106,970	149,249,385	6,161	404	7	41.28	65.44
Chemicals	39,691	61,525,574	757	29	0	12.30	0.00
Other industry	194,817	287,209,155	9,282	472	3	32.32	15.40
Tertiair	1,224,162	1,574,326,664	38,181	1509	20	24.25	16.34
Total	1,651,002	2,199,872,482	60,072	2969	40	27.31	24.23
Traffic accidents to and from work			12,784		28		

Source: Fonds vor Arbeidsongevallen (www.fao.fgov.be)

F1: Frequency of accidents (per million hours) OBO: Number of accidents with permanent injury

F2: Frequency of accidents (per million employees) DO: Number of lethal accidents

In Table 5.1 (CBS, 2006) we see the number of accidents for various industry groups in the Netherlands. Since the number of hours is not given, the accident rates are calculated per million employees. We can also see that there is a significant difference in the numbers reported by CBS and the numbers reported by RIVM (RIVM, 2006). This is remarkable, since both report the number of people killed on the job and one would expect these numbers to be well known. The difference amounts to an average difference in risk of some 30 per cent.

Table 5.2 gives the occupational accidents divided over the various main industry sectors in Belgium. If we look at these data more closely we can see that the number of accidents in the metal industry is much higher than that in the wood industry. But the number of accidents per hour worked is less. This means that probability of having an accident in the wood industry is higher than in the metal industry. The tertiair industry, which is everything industrial not listed explicitly, is much less risky than average.

We can see also that the construction industry in Belgium looks a lot riskier than the same industry in the Netherlands.

If we look at the number of people killed, the metal industry appears more risky than the wood industry. This shows that the result of a comparison depend on what you look at. As another example, let us compare the occupational accidents with accidents in general. Table 5.3 lists fatal accidents in a number of broad categories. We see that there are just over 3000 accidental deaths in the Netherlands each year, and that 50–60 per cent of these are the result of an accidental fall. The majority of the remainder are the result of road accidents.

Thus, it is interesting to look further into falls and investigate the circumstances. Here we see the limitations of our data. In almost half of the cases the circumstances of the fall simply are not known. Maybe the person was found dead and nobody saw the accident happen.

Of the other half, roughly a fifth are due to falls on steps and stairs. If we wanted to do something about falls, this is probably where we would look. But we do not know whether the larger number of falls is due to a large usage of stairs and steps, or whether the number of missions involving the use of steps and stairs when compared to the number of missions (to use the statisticians' term) involving snow and ice, could be an explanation for the difference in the number of lethal falls.

We can also compare the death rate of occupational accidents with other types of accidents in another way, as shown in Table 5.4. This table shows that an average Dutchman is seven times more likely to be killed on the road than to be killed in the course of his or her occupation. The probability of being killed at work is about ten times the probability of being killed by a stroke of lightning.

Table 5.3 Numbers of casualties for selected accident classes in the Netherlands, 2004

Year	Accidents							
	Total Accidents	Transport Accidents			Accidental Falls	Accidental Drownings	Accidental Poisonings	Other Accidents
		Total Transport Accidents	Road Accidents	Other Transport Accidents				
1994	3,430	1,314	1,261	53	1,550	104	60	402
1995	3,385	1,292	1,241	51	1,542	91	49	411
1996	3,372	1,314	1,198	116	1,605	71	86	296
1997	3,237	1,196	1,151	45	1,542	94	102	303
1998	3,059	1,090	1,044	46	1,540	83	72	274
1999	3,336	1,154	1,105	49	1,736	91	95	260
2000	3,345	1,123	1,085	38	1,675	107	123	317
2001	3,630	1,030	996	34	2,084	85	124	307
2002	3,364	1,055	1,001	54	1,830	115	128	236
2003	3,496	1,087	1,033	54	1,957	87	123	242
2004	3,228	878	836	42	1,839	96	162	253
2005	3,280	802	760	42	1,961	93	132	292

Source: CBS Statline (222.cbs.nl)

Table 5.4 Mortality of certain activities in the Netherlands

Activity	Mortality (per year)
Smoking	5.00E-03
Traffic (2007)	4.66E-05
Occupation	6.38E-06
Lightning	5.00E-07
Bee-sting	2.00E-07
Flood	1.00E-07
Falling aircraft	2.00E-08
Chemical industry	6.00E-09

Comparing risk

As I have argued before, risk is a matter of consequence and probability. Therefore, to compare the risks of activities, we have to look at the combination of consequence and probability. When the consequence is the same, we can simply compare the probabilities.

Figure 5.1 Probability of death per year of activity for workers in the Netherlands

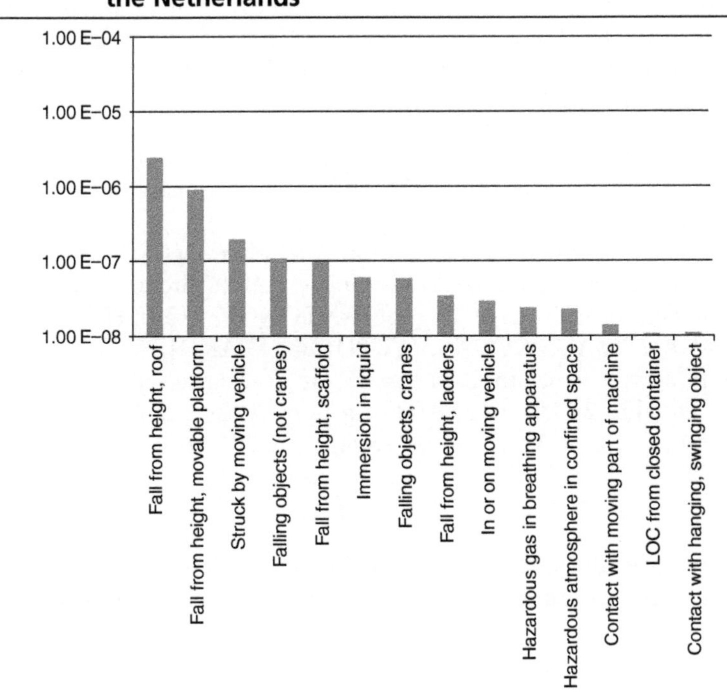

Figure 5.1 shows the probability of death for a number of work activities in the Netherlands. Now we can compare the risk of working on scaffolding with the risk of working on a ladder. We see that for the same amount of time, scaffolding has a higher death toll than ladders.

This does not necessarily mean that replacing scaffolding by ladders would make the risk lower. To make sure that such is the case, we should not only have to look at the same consequence (death) and the same period (in this case, a year), but also have to make other parameters equal, such as the height at which the work is being carried out and the type of work (such as cleaning windows or drilling holes in a wall). Only then could we truly compare like with like as Arnauld intended.

AVIATION

In accordance with Annex 13 – Aircraft Accident Investigation, states report to the International Civil Aviation Organization (ICAO) information on all aircraft accidents that involve aircraft of a maximum certificated take-off mass of over 2,250 kilograms. The organization also gathers information on aircraft incidents considered important for safety and accident prevention (International Civil Aviation Organization, 2000). These data are stored in the ADREP (Accident Data Reports) database.

The total number of occurrences in the database, from 02 January 1970 to 20 May 2007, is 34,912. One can find accidents of interest by putting a 'query' to the database. For European and North American commercial air transport, a query could look like this:

> ALL from 1990 not russian COMMERCIAL with mass groups and unrestricted
>
> Find all Occurrences where { Local date. after 31/12/1989 } and { Aircraft category. equal to Fixed wing } and { Aircraft manufacturer/model. doesn't have any of ANTONOV, ILYUSHIN, LET AERONAUTICAL WORKS, TUPOLEV, YAKOLEV } and { Operation type. equal to Commercial Air Transport } and { [Mass group. equal to 27 001 to 272 000 Kg] or [Mass group. equal to > 272 000 Kg] or [Mass group. equal to 5 701 to 27 000 Kg] } COUNT 8665

This query gives all the accidents of fixed-wing aircraft of Western make and heavier than 27 tonnes in commercial aviation from 1990 to 2007.

There are 8,969 occurrences of this sort. One could divide these into nine categories:

- abrupt manoeuvre
- cabin environment (fire, O2)
- uncontrolled collision with the ground
- controlled flight into terrain
- forced landing
- mid-air collision
- collision on the ground
- structural overload
- fire/explosion.

The relative frequency of these categories is given in Figure 5.2. It can be seen, for instance, that controlled flight into terrain (CFIT) is the second largest category, being the cause of a quarter of all accidents.

CFIT

A controlled flight into terrain (CFIT) is an accident in which the crew flies an otherwise normal aircraft into the ground. One of the most common reasons for this kind of accident is that the pilots have lost their spatial orientation. This could be on landing (perhaps because they think the runway is much closer than it actually is) or in mountainous areas (possibly because the pilots think they are much higher above the ground than they actually are). In order to avoid such accidents, a system called the Ground Proximity Warning

Figure 5.2 Distribution of accidents among accident categories

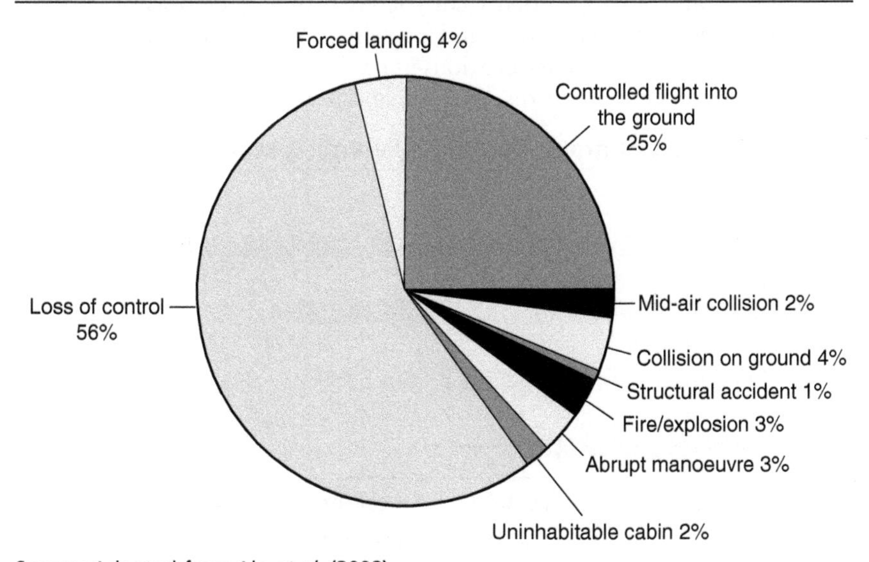

Source: Adapted from Ale *et al.* (2008)

System (GPWS) has been invented. The GPWS warns the pilots that they are close to the ground and advises – and in the latest version orders – them to cause the aircraft to climb. That such a system is effective can be seen from Figure 5.3. The frequency of such accidents has fallen considerably, and the better the GPWS system became, the more the frequency dropped.

Nevertheless, one would like to know whether the remaining accidents were sometimes caused by a failure of the GPWS system itself. Unfortunately, here is where the data reporting system fails utterly. Only in 4 per cent of all CFIT accident reports can information be found on the state of the GPWS: whether it was installed, whether it was operational, etc. This seems to be because it is illegal to take off without such a system. But it apparently does not occur to the keepers of the database that illegal acts are sometimes carried out regardless.

The latter finding holds an important lesson for users of any database. Always check whether the information in the database is complete and always check whether the information in the database conforms to the population you are looking at.

Trends

Another aspect of this analysis is that things change over time. At some time it is concluded that something needs to be done about certain accidents and then the probability changes – in the case of CFIT by more than two orders of magnitude. Clearly, then, the historical data are no longer valid. In the case of GPWS there is already enough history to draw such a conclusion. But what if you want to predict the beneficial effect of some change before the change is made? How can you be sure that what you are planning to do will help?

Here we enter the belief zone, and it is wise to keep in mind that there are often some serious implicit assumptions in predictions of this sort. Maybe the most common assumption is that everything else will remain the same.

Figure 5.3 Effect of ground proximity warning system

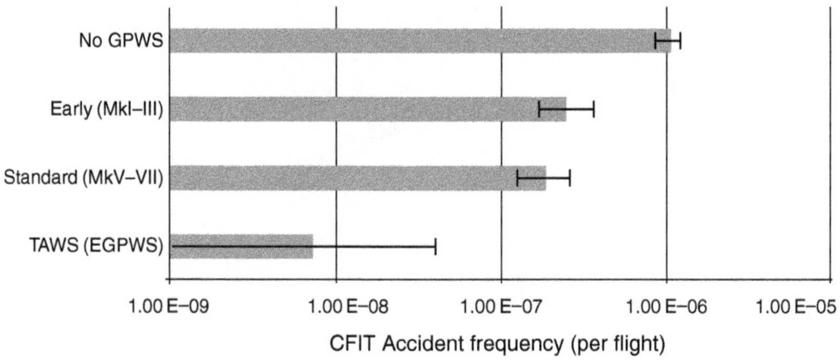

Source: Adapted from Spouge (2008)

Figure 5.4 The hockey stick curve

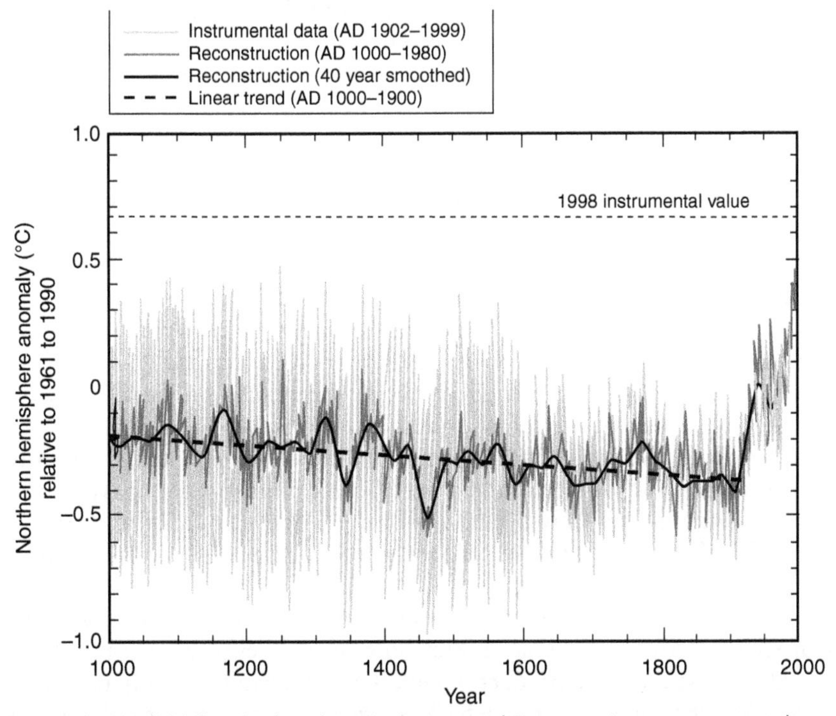

Source: (UNEP (2001) United Nations Environmental Program, Intergovernmental Panel on Climate Change (IPCC) Third Assessment Report, Climate Change 2001)

When seat belts were introduced in cars, it was predicted that the rate of death would decrease by 70 per cent. In reality it decreased by 40 per cent. The reason? Because in the prediction it was assumed that the behaviour of the drivers would not change other than that they would use their seat belts. But many drivers felt safer and drove at higher speeds. This counteracted the beneficial affect of the seat belt by almost half (NRC, 2004). Similar effects resulted from the introduction of other driver-assisting systems, such as automatic braking systems.

Global warming

If the stakes are really high, analyses of data can become the subject of deep scientific controversy. An example is the analysis of the behaviour of the temperature of the earth over time.

Figure 5.4 is the figure that is also in the Third Analysis Report of the Intergovernmental Panel on Climate Change (IPCC). This curve was introduced by Mann *et al.* in 1999. There has not been a continuous record of temperatures over the past 1,000 years. In the Netherlands, recording started in 1706. Therefore, Mann and colleagues used so-called proxy data, such as tree rings. If the curve is right – and most scientists now agree that it is – it means that the earth is warming up rapidly through anthropogenic causes.

This leads to some obvious questions:

- Is this bad?
- Can we do something about it?
- Are we willing to do something about it?

The current understanding is that it is indeed bad. It may not be so bad for those of us who live in the wealthy part of the world. But if you happen to live in the flood plains of Asia or in another famine-stricken part of the world, the change in climate looks as though it will make things much worse.

Since humankind created the problem, it might be possible to do something about it – if it is not too late. Here is one of the big unknowns, where beliefs dominate the consensus. Since we do not really understand how the climate system works, we really do not know whether we can undo what we apparently have done. But suppose that we could, albeit at an enormous price, then the cost would hit the United States, which uses a quarter of all energy despite having only 5 per cent of the world's population (http(1)), the hardest. So, it took decades before the world hesitantly agreed to reduce energy consumption.

We might think that these figures prove that the United States is wasteful with energy. However, if you look at how much energy is used per dollar earned in gross domestic product (GDP), the Netherlands actually uses more energy than the United States – which just reinforces the message about how careful one must be with numbers, data and the interpretation thereof.

The large interests at stake cannot but initiate continuous debate about how valid are the underlying assumptions and the methods used to estimate the history of the temperature of the earth, and although most scientists agree that the earth is indeed warming up and that the warming is the result of human actions, there also are fierce sceptics.

The risk of global warming

The potential effects of global warming and climate change may be disastrous. This, however, is not a risk of the kind we have been dealing with in the previous chapters. If it is said that the risk of climate change lies in sea-level rise and increased extreme weather, the probability does not mean that in, say, 90 per cent of scenarios in which the climate changes, sea level would rise. Probability in this context is the strictly Bayesian degree of belief. Experts are almost certain that sea level will rise but there is some doubt.

The problem here is obvious. There is no way of knowing for sure. There is no way of doing an experiment. There is only one earth and we cannot go back in time. This is what is called epistemological uncertainty: there is no way to know.

This is different from not knowing what number will come up on the next throw of a dice. That is called aleatory uncertainty and is caused by the fact that the outcome is probabilistic. But in principle the probability distribution can be known, or determined by experiment. The experiment may be impractical. It is impractical to perform say 50 controlled releases of 50 tonnes of LPG in an inner city to determine the precise probability distribution of the number of victims for such a release. But it is not impossible, and therefore the risk of LPG is much more a frequentist type of risk and amenable to quantification. What the 'risks' are of global warming is much more a matter of belief, albeit belief founded on present knowledge.

When experiments are impractical, or there is no way to really know, it is still possible to make an estimate. This we will discuss in the next section. But remember, the difference between not knowing and not knowable is fundamental, and we should keep that in mind when we are going to ask experts.

HUMAN ERROR PROBABILITY

A library of literature has been written about understanding human error (Dekker, 2006). Understanding the role of the human being and the reasons behind human error does not mean that we can also estimate or even quantify the probability of an error under certain circumstances.

In the CFIT example (p. 91) we have seen that sometimes humans are an essential part of the control loop, and the probability of failure has to be known in order to be able to estimate the probability of an accident. In other systems, such as nuclear power plants, the role of the human is more that of a back-up safety system. In that case it is important to know what the probability is that an operator will do the right thing in the rare case that he or she actually has to do something.

The only way to establish this probability is by observation. There only is a limited amount of factual information. Swain and Guttman published a handbook for human error probability in nuclear power plants in 1983, and much work has been done on car drivers (Houtenbos, 2008), but in industries such as aviation there is hardly any quantitative information available (Lin *et al.*, 2008). Given the importance attributed to human error and the money invested in reducing the contribution of human error to accident causation, this in itself is surprising.

However, we have to accept that knowledge of how our brain works is limited, and, since most of our processing is done by our subconscious, our brain is very difficult to influence (Dijksterhuis, 2008). This obviously limits the accuracy with which we can establish the probability of a mishap when and if the role of human action or inaction is significant in the causal chain of the accident.

In that light, it would probably be wise to eliminate human action in technical systems as much as possible, as the origin of Murphy's law also indicates, or, as Heinlein (1973) put it, 'Never underestimate the power of human stupidity.' However, the problem here is that often the creativity of the human brain also prevents accidents and leads to new discoveries.

As promised, the next section is about obtaining estimates without data: asking an expert.

NO DATA BUT STILL KNOWLEDGE

When data fail, we may have to resort to asking the opinion or judgement of experts. There are methods developed to do this in an organized and critical process (Cooke and Goossens, 2000). This process takes account of the differences in the level of competence of the available experts and also takes account of the perceptions these experts may have concerning their own expertise and that of others. The process calls for a group of experts to answer real and seed questions. The seed questions are used to assess the quality of the expert. Each time the expert is asked to give a 'value' of a parameter and an estimate about his or her uncertainty in this estimate. In this way it can be assessed how 'right' the expert is and how 'sure'. An expert who is right and sure is preferred over an expert who is right but unsure, and obviously an expert who is wrong but sure that he or she is right is the least desirable.

> 'I cannot imagine any circumstance in which a ship would perish. Modern ship building is past that stage.' This was said in 1907 by Captain Edward J. Smith, who worked for the White Star Line. He would later become captain of the Titanic, which sank in 1912.

However, in many cases the number of available experts is not even enough for such an approach to be possible. In that case the opinion or judgement may to a great extent be guesswork (Cerf and Navasky, 1984). As experts are often working for companies or groups with a certain stake at the decision, the possibility cannot be excluded that the judgements of these experts are coloured and framed by their working environment. The Health and Safety Executive (HSE) in the United Kingdom has found ways of dealing with these expert biases without losing the opportunity to employ what often is the best available knowledge and advice (HSE, 2000; House of Commons, 2006). The HSE selects experts from different groups of stakeholders and organizes discussions between these experts. The HSE then weighs the arguments. This involves a lot of judgement, but the whole procedure is referred to as expert judgement, so this is not really a problem.

The difference between the procedure of Cooke and Goossens and the procedure of the HSE is that the Cooke and Goossens procedure demands an assessment of the level of expertise of the experts. This is made on the basis of existing knowledge, and therefore the explicit demand is that there is knowledge that is relevant for the problem at hand. As we have seen before, there are problems for which there really is no existing knowledge.

Social Construct

In circumstances where the potential consequences of activities are strictly estimates and the probabilities are degrees of belief, the resulting assessment of the risk is called a social construct. This means that the majority of parties involved are able to agree what the consequences are likely to be. We have already seen the example of global warming. We cannot really know, and therefore whatever the risk is said to be is an agreed value.

A similar situation can be found in the use of nuclear power, where the period of time for which the nuclear waste has to be stored and watched exceeds the period for which humankind has any sort of historic knowledge, so again, whether this storage will be successful and the storage facilities undamaged over such a timescale is essentially a matter of belief.

We should keep in mind, though, that the 10,000 years or so involved in the storage of nuclear waste, and the corresponding statement that it is very unlikely that negative consequences will arise from that storage, is of a completely different nature from the statement that the probability of failure of some technical system is 1 in 10,000 per year. In the latter case the failure may occur tomorrow, and if we were to have 10,000 such systems, there would be such a failure on average every year, and the probability of not seeing a failure in a year would be less than 1 in 4, the probability of not seeing a failure in two years would be less than 10 per cent and the probability of not seeing a failure in four years would be less than 1 per cent. In the case of technological systems such as the chemical industry and in risks such as those relating to occupations, the risk can be quantified on the basis of observations and laws of nature that for the most part are understood.

You do not know what you do not know

Niels Bohr, the Danish quantum physicist, seems to have said, 'Prediction is difficult, especially for the future.' Quantum physics is all about probability, and Bohr took this probability as knowledge. What he meant was that even if we understand a system and its probabilistic risk now, we have to realize that if we extend that knowledge to future behaviour and risk, we have to assume that everything we did not include in our analysis of risk will remain constant. As the world is evolving, that is a strong assumption, and it is therefore wise to contemplate carefully whether it is a reasonable one. For dice used today and tomorrow that may be true. For dice used continuously for the next 10,000 years, that is probably not true. Normal dice will be worn down to a spherical shape by then (although a dice made from a diamond might not be). In addition, in the future things may happen that we cannot even imagine now. Who would have predicted the problem of SPAM? We just don't know what we don't know.

Finally, therefore, as regards this chapter it should be understood that a risk analysis of a chemical factory or of an aircraft is an analysis of what the risk is now, today, and what that same risk would be – now, today – if we were to change that system in some way. When the discussion is about global warming or the storage of nuclear waste, the discussion is about the future – the imminent future, as many scientists believe is the case for climate change, and the distant future, as many scientists believe is the case for nuclear waste.

If we are uncertain about what problem we are dealing with, a good question to ask ourselves is: can it happen tomorrow? If it can, it is usually a probability-type problem and amenable to a technical sort of analysis. If it cannot, then it is usually an uncertainty problem, which may or may not be solved by further analyses and observation.

Another way to distinguish between the two is the answer to the question whether what is uncertain is WHEN or IF. If we throw dice, sooner or later we will throw a six; what we do not know is when. These sorts of problems can be dealt with through analysis. Whether the sea will rise 10 cm, 1 m or 8 m we just do not know. So, either we will find out and we will know, or we will not know. If we do not know, it is NOT a matter of probabilities, but a matter of judgement or adventure or recklessness.

This difference has profound implications for the way decisions can be made. Unfortunately, not understanding this difference can lead to unfounded conclusions that also can have far-reaching consequences. This we will discuss in the next chapter.

6 Risk and decision-making

TAKING RISKS

> We Athenians, in our persons, take our decisions on policy or submit them to proper discussion ... the worst thing is to rush into action before the consequences have been properly debated. And this is another point where we differ from other people. We are capable at the same time of taking risks and estimating them beforehand. Others are brave out of ignorance; and, when they stop to think, they begin to fear. But the man who can most truly be accounted brave is he who best knows the meaning of what is sweet in life and of what is terrible, and then goes out undeterred to meet what is to come.
>
> (Thucydides, 431BC)

The concept of decisions on danger and risk is not new. And the concept that informed decisions are better than going for broke is not new, either. In the previous chapters we have seen how we can assemble the information that is needed to make a risk-informed decision. We also have seen that there are different sorts of uncertainty. In this chapter we will see how we can make decisions, depending on the sort of risk and the sort of uncertainty that is involved.

We prefer to die in traffic

To illustrate this point, let us refer again at the table of probabilities that we saw in Chapter 5, Table 5.4.

We can see that the probability of dying in traffic is about 10,000 times greater than the probability of dying as a consequence of an accident in a chemical factory. Yet society accepts the current state of affairs regarding traffic safety yet demands ever more stringent measures to ensure the safety of chemical operations (EU, 1982). Such demands are among the factors that led to the end of regular transportation of chlorine by rail through the Netherlands on 11 August 2006.

Apparently, more than just probability and consequences have to be taken into account when looking at risks. The same point was noted by people who had to deal with the risks of nuclear power generation. Although, as they said, the risk of a fatal accident with a nuclear power station was much smaller than the risks associated with traffic accidents or smoking, that did not make the risk necessarily more acceptable. A continuous problem in that discussion was the argument by the public that a nuclear power station could explode like a nuclear bomb, while the experts asserted that such was physically impossible. This led Kaplan and Garrick (1981) to the introduction of risk triplets.

The risk triplet

In Kaplan and Garrick's view, risk consists of three elements: the scenario, the probability and the consequence:

$$R = f(s, p, c)$$

What they also want to emphasize is that it is not sufficient to state that a certain consequence has a certain probability; one at least should have a scenario, a plausible story, of how the consequence could actually come about.

Kaplan and Garrick came to this notion because they were primarily dealing with the risks of nuclear power generation. The ultimate doomsday scenario of a nuclear power station is a mass nuclear explosion of the whole of the reactor, like an atomic bomb. In most, if not all, nuclear power stations, such a scenario is physically impossible. However, in many discussions about risk one would set the probability of such a physically impossible event not to zero, but to a very small number. This was (in the 1960s and 1970s) a common way of dealing with epistemological uncertainty. Even if science proclaimed that something was impossible, that was not necessarily the ultimate truth. There could always be some mechanism, not known to science, that would make an impossible scenario possible after all. Given the enormous consequences of such scenarios, taking a small number rather than zero made an enormous difference for the risk. Kaplan and Garrick argued that one should have at least one, plausible, scenario before one could set the probability to a non-zero number. (But even then we should be careful. One in a billion (10^{-9}) per year may look a small number, but if this is the frequency of wiping out a significant part of the world's population, it is in fact double the risk of human extinction as a result of a supernova in the general neighbourhood of the Earth, which is estimated at between 10^{-9} and 10^{-10} per year.) In addition, the description of the scenario then could be used to find ways of reducing the probability or the consequences or both by doing something about this scenario.

These days the description of the scenario is the basis for what is generally known as 'risk characterization'. The character of the risk is determined not only by consequence and probability, but also by a host of non-quantifiable entities that together play a role in the decision-making process associated with accepting an activity that produces risk – usually as an unwanted side effect – and thus the risk. An example of these characteristics is given in Figure 6.1.

Figure 6.1 Risk characterization elements

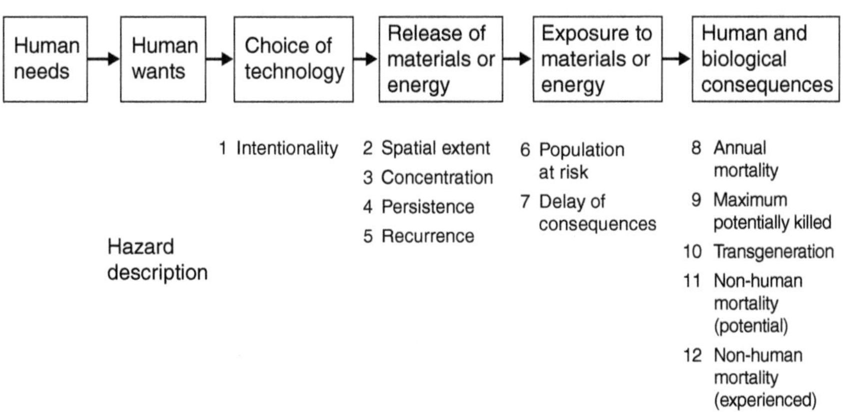

Source: TU-Delft

How risks emerge

As I pointed out in Chapter 1, unless we are 'thrill seekers', most of us rarely take risks for the sake of it. We accept risks as part of an activity we undertake. We undertake these activities because we have to or because we want to.

Some activities are necessary. We have to eat. Therefore, we have to acquire food and drink. In prehistoric times we would go hunting, in forests and on plains inhabited by creatures that would try and eat us before we could eat them.

The reason why we choose to undertake an activity or a particular technology is closely associated with our judgement as to whether the risk is worth taking, given how much the activity means to us. Therefore, people can differ substantially over the acceptability of risks that in terms of their (quantified) probability and consequence are equal or similar. This became probably most explicit and a matter of importance for energy policy, first in the United States and later all over the world, in the discussion about nuclear power generation – a debate, in which Kaplan and Garrick, whom we have come across already, also had a part. In this discussion the divide between those who accepted the risk of nuclear power – and judged the risk to be marginal – and those who did not accept the risk – and judged it to be large – coincided with the divide between experts and laypeople. This divide intrigued many social scientists and led to decades of studies on how people see and judge risks: risk perception.

Risk perception

The phenomenon that we judge risks differently depending on the kind of activity that generates the risk is also called risk perception. In part, these perceptions are driven by the way in which information is processed by our

Table 6.1 Probabilities of death and probabilities of winning certain lotteries

Activity	Winning a lottery	Probability (per year)
Smoking		$5*10^{-3}$
Traffic		$8*10^{-5}$
Lightning		$5*10^{-7}$
Bee-sting		$2*10^{-7}$
Flood		$1*10^{-7}$
	Staatsloterij	$1*10^{-7}$
	Bankgiroloterij	$4*10^{-8}$
	Lotto	$2*10^{-8}$
Falling aircraft		$2*10^{-8}$
	Postcodeloterij	$1*10^{-8}$
Chemical industry		$6*10^{-9}$
	Sponsorloterij	$3*10^{-12}$

brain. Table 6.1 shows the mortality of various activities. These figures apply to the Netherlands. From the table it can be seen that the probability of any Dutchman being killed by an accident in a chemical plant who is not an employee of the plant is six orders of magnitude smaller than the probability of dying of a smoking-induced illness (if he or she is a smoker). On the basis of these numbers a decision maker has a fair point when assuming that the probability that he or she will be confronted with a disaster in the chemical industry is remote and hardly probable – especially when one notes that the Netherlands in its present form is just over 200 years old. The first constitution was accepted in 1798.

Table 6.1 also gives the probabilities of winning the main prize for five of the nation's lotteries. One can see that the chance of winning the 'Sponsor Lottery' is three orders of magnitude smaller than that of being a victim of a chemical accident. Nevertheless, these lottery tickets are readily sold and there is regularly a winner. Apparently many people consider it highly unlikely but possible, or even probable, that they will win this lottery. This difference in appreciation of the numerical information is closely related to the psychosocial theories of risk perception. According to these theories, many factors shape the perception of risky activities (Vlek, 1996; Slovic, 1999; Sjoberg, 2000). The top ten of those most listed are:

- extent and probability of damage
- catastrophic potential
- involuntariness

- lack of equity
- uncontrollability
- lack of confidence
- new technology
- non-clarity about advantages
- familiarity with the victims
- harmful intent.

Slovic *et al.* (1978) combined these into a two-dimensional scale of 'awfulness' (Figure 6.2). The awfulness of an activity in this figure is determined by two parameters that embody the combination of the ten factors listed above. These two parameters are whether the risk or activity is well known and whether there is some dread associated with the consequences. Dread is the combination of factors such as disastrousness and delayed, not readily detectable effects.

Combining these factors with the mortality discussed earlier reinforces that people are more willing to accept a certain small loss than an uncertain large loss.

The way we learn

If disasters have such an effect on the way we judge risks, how come the attention paid to these risks tends to dwindle away after a while, letting disaster strike again (Ale, 2003)? In the Netherlands, some large-scale

Figure 6.2 Scale of awfulness

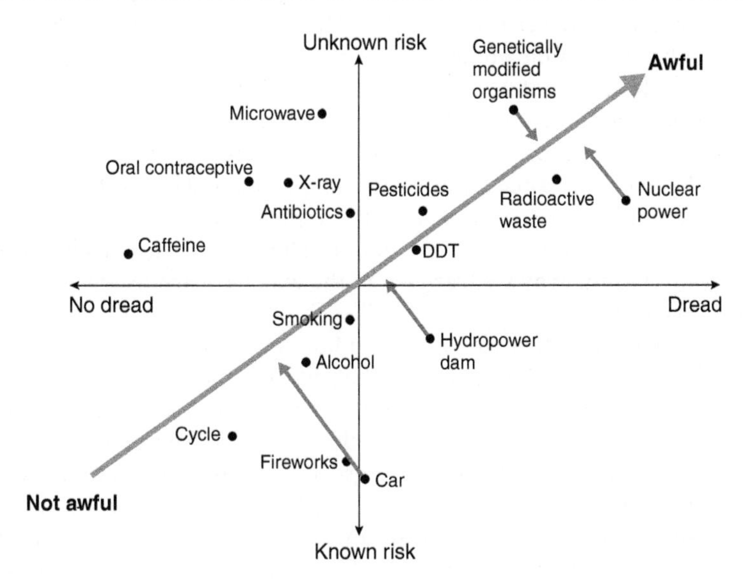

Source: Adapted from Slovic *et al.* (1978)

Figure 6.3 The 'big thunder' of Delft, 1654 (by an unknown artist)

accidents to do with explosive materials have occurred. In 1654 the centre of Delft was demolished by the explosion of a powder tower. This explosion (Figure 6.3), which could be heard 80 kilometres away, created the 'horse market', which is still an open space.

In 1807 a similar explosion took place when a barge laden with black powder exploded in the centre of Leiden. The van der Werf park today is still witness to this event. One hundred and fifty people were killed, among them 50 children whose school was demolished by the blast. On 13 May 2000 an explosion occurred in a fireworks storage and trading facility in Enschede, also in the Netherlands. Twenty-two people were killed and some 900 injured. The material damage was costed at approximately €400 million.

So, every 150 years or so a disaster happens because explosives that should not have been in a city were in one. The reason why this happens is closely related to the way we learn – and the way we learn used to serve our survival very well.

Long, long ago, when we saw that a lion had eaten our nephew, it was a good idea to learn and assume that that was what lions do: eat nephews. And furthermore, if the next time a lion passed it did not harm us, it was a good idea not to change our mind immediately about lions. Thus, our brain works such that information that strengthens existing ideas is more readily absorbed than information to the contrary (Reason, 1990). This works fine – except when the probability of an adverse effect is small and the time between accidents or disasters is long. The way we learn leads us to think that accidents can and will not happen. Because the probability of a large disaster is small, long periods of time may elapse after one disaster before another strikes. In this period the notion that improbable equals impossible is steadily reinforced, and thus the impetus that exists shortly after a disaster to do something about it disappears.

These disasters do not have to be large-scale events that affect society as a whole. The way we learn also affects the way we take decisions on a smaller scale. If a house in the neighbourhood is broken into, we become more cautious. We check whether the windows are closed before we leave and maybe even install better locks. But then, after a while, we leave the window open again if it is hot outside and perhaps even put the key back under the flowerpot.

Risk as societal construct

The notion that there is more to a decision on risk than probability and consequence has induced definitions of risk that encompass the other, more intangible aspects. In these definitions, risk is a construct of society. Society determines what risk means.

The problem here is that the definition of risk then starts to determine what is taken into account in the – political – decision-making process. Only those aspects that are part of the definition of risk are weighted. Therefore, the agenda of the discussion is set by the risk qualifier rather than by the institutions and organisations whose purpose is to discuss and decide, or even by society itself.

Thus, it seems a better idea to restrict the definition of risk to the combination of probability and consequence and leave the listing of all other aspects that may be important in the discussion open for debate. Many of these aspects fall under the headings listed on p. 102. But there are many others. The debate about nuclear power may be about the safety of the power plants and the storage of nuclear waste, but the debate is fed by opinions about the form of governmental organization that is necessary to provide sustainable control over the fate of the fuel and the waste. The debate about the risks of terrorist attacks and the countermeasures is less about the risk than about the invasion of privacy and the prolonged storage of private information. And often with good reason. The comprehensive records on population kept by the Dutch government helped in the planning of housing and social security, but also led to a high proportion of Jews being deported and killed in the Second World War.

Decisions can also be driven by personal ambition, by the desire to grow, by market competition, and the desire to Do Great Things.

The acceptability of risk is never an isolated issue. Risk comes about as a result of a decision about an activity, which may have many more aspects and qualities than risk alone. The risk analyst has to be aware of these arguments in the debate to understand what is done with his or her analysis, and the risk manager has to understand how the boundary conditions under which he or she operates come about.

Risk itself may be called a social construct. What is really a social construct is the definition of the problem to which the decision relates and the things we want to consider as relevant. Risk may be part of it, and uncertainty also.

Disaster aversion

Nevertheless, even in purely economic terms it sometimes makes sense to avoid a disaster. This is the case when we cannot afford the consequences of such a disaster. A typical – household – example is fire insurance. To take out fire insurance must on average be a bad deal. The insurance companies make money, and the only way they can do so is when on average those who are insured lose. However, an individual homeowner would be bankrupt if his or her house burned down, because the individual would not live long enough to make up the loss, and therefore cannot take the risk.

After the 9/11 attack, insurance companies all over the world issued letters to their clients stating that they would not pay for damage resulting from a terrorist attack that exceeded one billion euros in total, even if all clients had paid all their premiums in time.

This notion of disaster aversion leads to the presentation of risks in risk matrices in which risks are depicted in two dimensions, probability and consequence, without combining them into a single number.

Risk matrix

In risk matrices the two dimensions of risk – probability and consequences – are separated out and plotted against each other. Any combination of consequence and probability is a point in this two-dimensional space. Alternatively, the risk profile of any activity can be plotted as a so-called complementary cumulative distribution function (CCDF). In such a curve the probability of exceeding certain consequences is given as a function of these consequences (Figure 6.4).

The plot area can be divided into three areas: acceptable (green), conditionally acceptable (yellow) and unacceptable (red). Whenever the risk is not in the acceptable area, measures have to be taken, or at least contemplated. Of particular interest is the region in the lower right-hand corner of the matrix, where risks whose consequences cannot be borne are located. These risks have to be transferred, perhaps by insurance, or have to be eliminated – regardless of how low their probability is – as the consequences would lead to ruin. In practice, any consequence proves to be acceptable when the probability is sufficiently remote and the advantages to be gained by embarking on the risky activity are sufficiently large. Therefore, the dark grey or unacceptable

Figure 6.4 Risk matrix

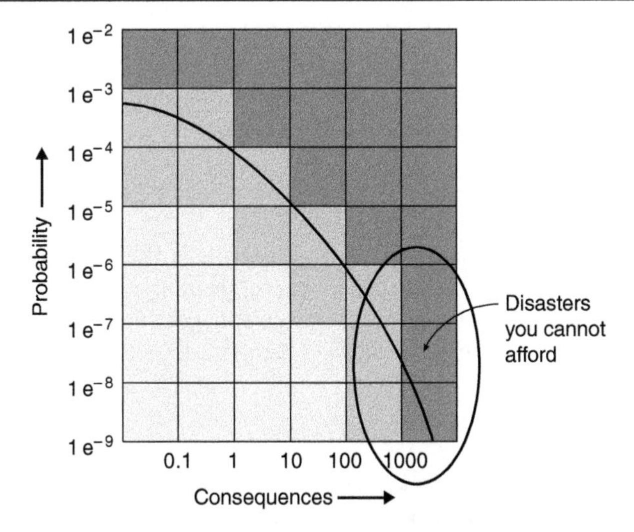

area is seldom demarcated by a vertical line. Rather, the limit is some sort of sloping line, as depicted by the curve in Figure 6.4

The use of risk matrices is no longer restricted to the chemical industry. Many applications are found in financial services and insurance (McCarthy and Flynn, 2004).

RISK METRICS

There are many ways by which risk can be expressed. In this context, 'expressed' means something different from 'characterize'. Characterizing is done in terms of large or small. Metrics are used to express the extent of a risk in a (semi-) quantitative sense – on an absolute scale, or in terms of how different risks compare.

As we have seen, risk is a combination of the – extent of the – consequences and the probability that they will occur. Thus, any metric for risk has to have these two dimensions, either combined or separately.

When one is making a risk analysis to support a decision, the choice of the metric itself can have an influence on the outcome of the decision-making process. Once the metric is chosen, only risks that can be expressed in this metric will be part of the decision making. This is also called framing. Everything in the frame will be considered, everything outside the frame will not. Before we discuss framing a little further, let us first look at some metrics. A number of metrics are widely used in the risk management field and these are discussed below.

Expectation value

The expectation value is also called the weighted average or the mean. This is the metric that often is referred to as chance multiplied by effect. C_i are the possible consequences of an action or activity and pi are the probabilities of Ci then the expectation value of the risk is

$$E(R) = \sum_i P_i \times C_i$$

If the consequences are people killed, the result is the average number of people killed per event. If the consequences are material loss, the result is usually expressed in dollars or euros or yen lost per event. This is the value of the risk that Arnaud in 1662 advised using in decisions about activities.

Frequencies and probabilities

Often it is more desirable to know the average loss rate than the average loss. For instance, regarding traffic safety one wants to know the number of casualties per year or the material damage per year. In that case, probability is replaced by frequency. Note that frequency and probability are two completely different entities, and numerically can be completely different too. As an example, the number of lethal car crashes per year in the Netherlands is approximately 1,000. This makes the probability of a car crash in the Netherlands in any given year equal to almost 1. A probability is always a number between 0 and 1. A frequency can be any positive number.

Individual risk

When considering safety, one is often interested in the probability that a particular individual will come to harm. This harm can take the form of a financial loss, injury or illness but also death. The probability that a person will come to a particular harm is called **individual risk.** Individual risk also is often expressed as a frequency or rate. The fatal accident rate is the probability of a worker being killed in a working year. This really is a frequency (number killed per year). The time here is not a complete clock-year (356 days, 24 hours) but more the some 1,500 hours that constitutes a working year. This difference is important. For instance, in a factory where production goes on 24 hours a day with three shifts, one job position actually has 3 (rather than 1) working year per year.

Localized risk

A variant of individual risk is localized risk (*plaatsgebonden risico* (PR) in Dutch). Localized risk is risk associated with a location. In spatial planning it is usually the frequency that a lethal level of some agent – chemical, a shock wave, heat radiation – is exceeded at that location. The frequency is usually expressed in units per year.

Group risk

Group risk or societal risk is the probability or frequency that a group of a certain size will be harmed – usually killed – simultaneously by the same event or accident. It is presented in the form of an FN curve. This sort of curve is really a so-called complementary cumulative distribution function (CCDF). In a CCDF, each point on the line or curve represents the probability that the extent of the consequence is equal to or larger than the point value. As both the consequence and the frequency may span several orders of magnitude, the FN curve or FN line usually is plotted on double logarithmic axes (Figure 6.5).

If the expectation value and the FN curve are calculated on the same basis (that is, the same definitions of frequencies and consequences), the surface area under the FN curve is equal to the expectation value. This is a useful property as sometimes this area is easier to calculate from a list of potential accidents.

As we have seen, the same concept can be used in finance and insurance, where the loss is now expressed in terms of money. A catastrophe is a loss that would make a company insolvent or cause an insurer not to be able to pay out the claims made on it. It has been said that the asbestos claims did more damage to the English nobility than Robespierre did to the French nobility. The underwriters of Lloyds are mainly wealthy English people who guarantee the insurance policies with their own personal capital. As a result of the asbestos claims, many had to sell off their estates to back up their signatures. I shall assume that readers know that Robespierre orchestrated the beheading of the French nobility in the eighteenth century.

Figure 6.5 FN curve and expectation value

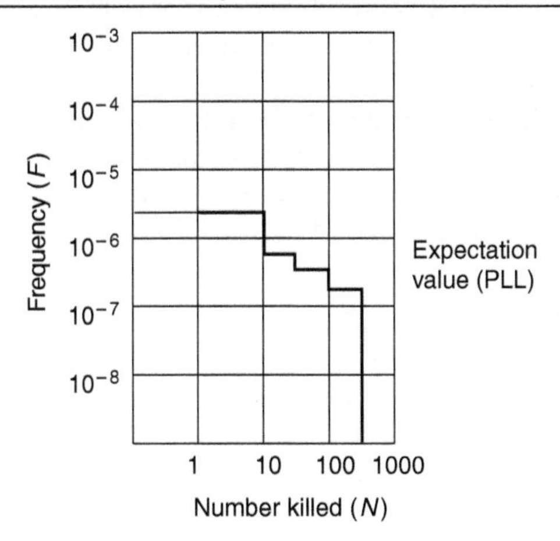

RISK ACCEPTABILITY AND CRITERIA

In all problems regarding safety, the ultimate question is, how safe is safe enough? More safety usually costs more money, and in any case there is no such thing as zero risk. If we ban or discontinue one activity that involves risk, we have to undertake another that produces another risk. Risk cannot be completely banned, as we are all mortals and we shall all die from some cause or another.

As we have seen, there are many factors that determine the acceptability of risk. Nevertheless, to execute a risk management policy or strategy the acceptability has to be determined and measures have to be implemented to ensure that the actual risk does not exceed the level of acceptability. There are a number of ways of doing so. One may set standards or one may define a method of dealing with the problem, such as cost–benefit analysis, or one may set out a principle to guide how the problem is dealt with. The ALARA (As Low As Reasonably Achievable) principle is an example; the precautionary principle is another.

Principles

There always has been a desire to reduce the risks associated with human activities. Even when we do dangerous things such as Formula 1 racing, we still reduce the risks as much as possible, although it would be safer not to race at all. The desire to limit the side effects of human actions led to the ALARA principle.

ALARA and ALARP

Risks, but also pollution and other side effects of activities, should be as low as reasonably achievable. In the United Kingdom this idea was formulated as the **ALARP principle:** As Low As Reasonably Practicable. This change of word indicates that achievability cannot only be a matter of theory, but must be possible in practice. For all practical purposes, however, ALARA and ALARP are the same.

The problem with this principle has been to define what is reasonable. The opinion about what is reasonable differs from country to country.

In the United Kingdom the question of what is reasonable was defined by Lord Justice Asquith in 1949 (NN, 1949):

> [A] computation must be made in which a quantum of risk is placed on one scale and the sacrifice, whether in money, time or trouble, involved in the measures necessary to avert the risk is placed in the other. If it be shown that there is a gross disproportion between them, the risk being insignificant in relation to the sacrifice, the person upon whom the duty is laid discharges the burden of proving that compliance was not reasonably practicable.

In any case it is important to realize that the above constitutes one of the basic principles of the common law tradition: what is not explicitly allowed is forbidden unless it can be justified, where necessary in court. When such a guiding principle is combined with the obligation of the hazard producer to demonstrate that he or she has done what is reasonable and practicable, this can be a powerful risk-reducing policy. Certainly it can mean that design and operation have to be instituted such that standards are not violated if such violation could have been avoided easily. Whether this is indeed the case will depend largely on the willingness and the capacity of the regulator to implement these principles.

However, in a increasingly competitive environment, such as is developing in the United States and the European Union, any cost is damaging to an enterprise's competitive edge. Any expense made to lower an emission or a risk below any available standard can and will be called unreasonable or excessive. This is also the opinion of the courts in a country such as the Netherlands, where it is considered unreasonable if the costs exceed the benefit even slightly, and where the courts are of the opinion that cost-bearing risk-reducing measures cannot be demanded unless they are required by law, which is more in the tradition of Napoleonic law, where everything is allowed that is not explicitly forbidden (Ale, 2005). In that case the policy defaults to a standards- rather than a direction-driven policy.

A further problem may lie in the uncertainty that exists about how safe is safe enough or how clean is clean enough. This is of special interest to multinational companies. When one can reach a certain level of safety in one country, why not have the same level in another? It is therefore cost-effective to negotiate as little additional safety as possible beyond what is needed for continued operation in every country in which the company has a presence.

The Precautionary Principle

The precautionary principle is that an activity should not be embarked on if there are risks. This seems a laudable principle. However, it too comes with a price tag. There is no a priori reason to assume that the new risks are worse than the existing ones and that the risks will not prove to be worth the benefits (Jongejan *et al.*, 2006).

However, in many instances where the precautionary principle is invoked, we are no longer confronted with calculable risks. Often these problems transgress the boundary between probabilistic uncertainty and epistemological uncertainty. We have seen that we then enter the realm of guesswork – hopefully, educated guesswork – but when there is no reason to assume that the past can be extrapolated to the future, there also are no guarantees, either for better or for worse.

COST–BENEFIT ANALYSIS

Cost–benefit balancing starts with the notion that risks are not taken for no reason. Risks are a side effect of benefits that we want to obtain. These benefits are to some extent balanced by the costs incurred by the risk. As long as the benefits outweigh the risks, the risk is worth taking. If the benefits are larger than the risks, some additional expenditure may be taken to reduce the risk, until risks and benefits balance. There are great advantages to this approach in the context of a market economy because it takes safety from being a vague ethical concept to being a good with a price worth paying. There are also significant problems with the approach, however.

The problem here is that the risks have to be expressed in the same units as the benefits, which is easy if the potential losses are material. However, if the potential losses are human health or life, the question of valuation becomes far more difficult to answer. There are ethical and personal – subjective – valuations at stake even more than with the perception of risk, which also involves subjective valuations.

Since cost–benefit analysis is one of the most widely used decision support tools, many attempts have been made to assess what in general the public and decision makers are prepared to spend on saving lives. These attempts have led to widely varying results (Morrall, 1986, 1992; Tengs *et al.*, 1995). And as recently as at the ASME international conference in Annaheim in 2004 it became apparent that in a proposed framework of making choices to prioritize the defence options to protect critical infrastructures against outside threats, human lives could not be put in what was otherwise a completely monetary framework (Nickell *et al.*, 2004). The problems are associated with the valuation of environmental quality, the value of human life and the valuation of risk abatement costs.

Benefit

As far as the value of human life is concerned, a continuous stream of efforts have been made to deduce a number. All of them have had an ethical drawback of some kind (NN, 1995). This leads to tables such as Table 6.2, a preliminary catalogue of the cost of saving life.

Recent literature indicates that these estimates can be very misleading because people manage their risk much more intelligently than by just looking at the one risk under consideration. They look at their total portfolio of risks and may compensate for an increased level of risk by taking risk-reducing measures somewhere else. For instance, rather than always sticking to a moderately healthy diet, they may decide to compensate for a heavy meal one day by a day of fasting the next.

If earning power is used (Morrall, 1986), the question arises as to how to deal with unemployed people. If the number of life-years lost is used, the

Table 6.2 A preliminary catalogue of the cost of saving life (US$ 1983)

	Measure	Approx CSX
1	Nuclear power plant hydrogen recombiners	3,000,000,000
2	1981 Automotive CO standard	1,000,000,000
3	OSHA benzene regulations (USA)	300,000,000
4	Apartment building regulation (UK)	100,000,000
5	Pharmaceutical industry safety measures (UK)	50,000,000
6	Grounding of DC-10 aircraft 1979	30,000,000
7	Lighting of all X-ways (USA)	10,000,000
8	Nuclear waste treatment (USA)	10,000,000
9	SO_2 scrubbers (least effective case)	10,000,000
10	General nuclear safety UK	10,000,000
11	Safety of nuclear operatives	5,000,000
12	Nuclear reactor containment (USA)	5,000,000
13	OSMA coke fume regulations (USA)	4,500,000
14	EPA vinyl chloride regulations (USA)	4,000,000
15	EPA drinking water regulations (USA)	2,500,000
16	Nuclear reactor diesel sets (USA)	1,000,000
17	Home kidney dialysis sets	640,000
18	Kidney dialysis	532,000
19	CPSC upholstery flammability standards	500,000
20	Highway rescue cars (USA)	420,000
21	Sulphur stack scrubbers	320,000
22	Auto air bags	320,000
23	Auto safety improvements, 1966–1969 (USA)	320,000
24	Reduced infant mortality (Canada)	210,000
25	Kidney dialysis treatment units	200,000
26	Kindney transplant and dialysis	148,000
27	Highway safety programmes (USA)	140,000
28	Mandatory lap–shoulder belts in cars	112,000
29	SO_2 scrubbers (most effective case)	100,000
30	Nuclear reactor ECCS (USA)	100,000
31	Life rafts in aircraft (UK)	85,000
32	Car seat belts (USA)	80,000

Table 6.2 A preliminary catalogue of the cost of saving life (US$ 1983) *continued*

	Measure	Approx CSX
33	Home smoke detectors (USA)	80,000
34	Smoke detectors (USA)	65,000
35	Agricultural practices (UK) pre 1969	50,000
36	Tuberculosis control (USA)	42,000
37	Mobile cardiac emergency units (USA)	30,000
38	Cancer screening programmes (USA)	30,000
39	Early lung cancer detection	16,000

Source: Morrall (1986) A review of the record, regulation, vol 10 no 2.

question becomes how to deal with cases that place the elderly or the young in specific hazardous situations. The notion of quality-adjusted life-years (QALY) raises the question of handicapped or challenged people as against the healthy. In terms of expenditure per life-year the numbers range from 0 to 99 x 109 US dollars. The policy value of a human life seems to gravitate to approximately US$ 7 million. This is equivalent to $200,000 per life-year. If this indeed proves to be the value of a human life-year, it does not seem to be a large amount, especially when multimillion revenues are at stake.

According to Button (1993), the official value of a human life varies from €12,000 to €2.35 million and the market value between €97,000 and €630,000. In more recent literature, the value for a human life to be used in cost–benefit analyses is called the value of saving a statistical life rather than the value of a statistical life (VOSL). The argument made is that the value of life is not capable of being measured. Nevertheless, in any cost–benefit decision the bottom line is that the value used is the value for letting that life continue in that decision. And in decisions where statistical lives are lost, even when that loss is called tolerable (HSE, 2001), the activity and the risk are accepted, and decision makers would fare better in the public debate if they did not hide this behind words. HSE (2001) uses a value of £1 million in its advice on risky activities. The UK Department of Transport and the Regions uses £3.6 million in transportation risk studies (Chilton *et al.*, 1998). In an evaluation that is more recent than the one performed by Tengs *et al.* (1995), Viscusi and Aldy (2003) derive a value of US$4 million.

Even more difficult is the valuation of injuries and of environmental damage. A comprehensive description is given in the Eterne (NN, 1995) report, but even in that report the results remain inconclusive.

Cost

There are also great difficulties in assessing the actual value of the risk reduction costs. A seemingly expensive measure such as the desulphurization of residual oil actually raised revenue. Many of the expenses made in connection with safety also result in increased reliability of production and lower costs, resulting from less downtime and fewer accidents. In fact, in the study performed by Pikaar and Seaman (1995) for the Dutch Ministry of Housing, Physical Planning and Environment it appeared that most industries did not consider it worth their while to register the costs of these measures. This is consistent with the findings of Tengs *et al.* (1995) of the low costs of measures related to the prevention of accidents.

So, the methodology for cost–benefit analysis itself is well established and works well when only material loss and gain are involved, such as in finance. However, there currently is not yet a clear understanding of how to use cost–benefit analysis in an organized way in risk management when these risks go beyond financial risks. What to put in monetary terms against what otherwise is called 'imponderable' is still the subject of considerable debate.

As I have said, especially when human life and health are at stake, the discussion is more about values than about costs. But in real life, decisions are often much more cynical than that (Evans, 1996).

The discussion about Automatic Train Protection (ATP) systems in the United Kingdom is illustrative in this respect. After Lord Hidden (1989), in his report on the Clapham Junction accident, advised that ATP be introduced on the British Rail network, British Rail issued a report (British Railways Board, 1994) in which it showed that, given the value of a statistical life used in the United Kingdom, ATP was disproportionally expensive. This conclusion was endorsed by the Department of Transport (1995). After the crash at Ladbroke Grove, Lord Cullen (2001) concluded that that decision was not unreasonable, but that the introduction of ATP was advisable regardless of these costs. This in turn led Evans (2005) to the conclusion that the money needed could be better spent on road safety, illustrating again that monetary cost–benefit evaluations seldom dominate the final decision.

Given these experiences, the weighting of human lives in an economic framework has to be left to the decision makers, and any analysis that quantifies human losses in terms of money has to be explicit about the valuation.

SETTING STANDARDS

Norms and standards have a proven use in the technical community. Without the standard for the base of a lighting bulb, street lighting would not have been possible. In general, standards and norms make it possible to have such things as spare and replacement parts without the necessity to manufacture each object when and where it is needed. It also makes technologies interchangeable and transferable.

Standards for quality of consumer goods protect the consumer against faulty or dangerous products. They prevent producers from making quick money on low-quality merchandise and thus level the playing field for the players in the market: the providers and the consumers.

Standards in legislation also make decision making predictable, allowing manufacturers to design products that meet regulatory requirements. They also make the demands uniform in a particular country, or in Europe as a whole.

These technical norms and standards have all quite a strict, obligatory character. It is not enough merely to strive to attain them. The norms should be met. Imagine trying to fix a flat tyre on a roadside and it turned out that the manufacturer of the spare tyre had tried hard to meet the specifications but failed!

Risk criteria

The strict character of norms is their advantage in a technical or operational situation, where routine decisions and processes prevail. They are less suitable in circumstances where tailor-made solutions are required. Whether it is best to regulate by norms or by indicating the general direction of the required development depends on the characteristics of the problem to be regulated. One way of controlling risk therefore is to set standards. If these are expressed in terms of risk, they are called risk criteria. As the factors that influence the judgement concerning a risky activity are different for differing activities, it cannot be expected that a single set of risk criteria will be applicable to all activities. Even so, a policy may look more organized if the set of applicable criteria is kept small.

On the other hand, it is argued that these factors make it impossible to set general standards, as every situation and every activity is different. A more extreme stance argues that risk is a social construct rather than something that in principle can be determined scientifically. In this view there are so many subjective choices made in risk analyses that they cannot be called objective science at all (van Asselt, 2000). Scientists are just other laypeople. Their judgement is influenced by the same factors, but in addition they let their science be influenced by their political judgements. As we have seen, it depends on the sort of activity and the sort of risk we are looking at whether this is a defensible viewpoint.

It is no surprise that the more objectivist risk analysts argue that scientific judgements and political judgements are not the same thing and that objective quantification of risk is a scientific exercise. Indeed, such objectivity is necessary to make cost–benefit-based decisions. These people argue that the value of the risk should be as objective as the – monetary – value of potential risk-reducing measures (Tengs *et al.*, 1995).

Even when we accept that standards for the tolerability of acceptability of risk can be set, these still can differ for different classes of risk.

External safety

For external safety, several countries have set limits to the acceptability of risk, either by law or via decrees. Let us use the Netherlands as an example of how these limits come about. In the Netherlands, limits have been set for localized risk and group risk, the latter being advisory (Figure 6.6).

These risk limits evolved from decisions that had to be taken as a result of economic developments and accidents. Regional and local authorities as well as industry asked for guidance regarding the acceptability of risk. The basis for this guidance was found in documents and decisions taken earlier. An important baseline was found in decisions made regarding the sea defences of the Netherlands. In 1953 a large part of the south-west of the Netherlands was flooded as a result of a combination of heavy storms, high tides and insufficient strength and maintenance of the dyke system. Almost 2,000 people lost their lives and the material damage was enormous, especially as the Netherlands was still recovering from the Second World War. The country embarked on a project to strengthen its sea defences, including drastically shortening the coastline by damming off all but one of the major estuaries of the Rhine–Maas delta. The design criteria were determined on the basis of a proposal of the so-called Delta Committee, which proposed that the dykes should be so high that the sea would reach the top only once every 10,000 years (Delta commissie, 1960). The probability of the dyke collapsing was lower by a factor of 10. The probability of drowning was lower by another factor of 10, so that the recommendation of the Delta Committee implied an individual risk of drowning in the areas at risk of one in a million per year. This recommendation was subsequently converted into law.

Figure 6.6 Risk limits in the Netherlands

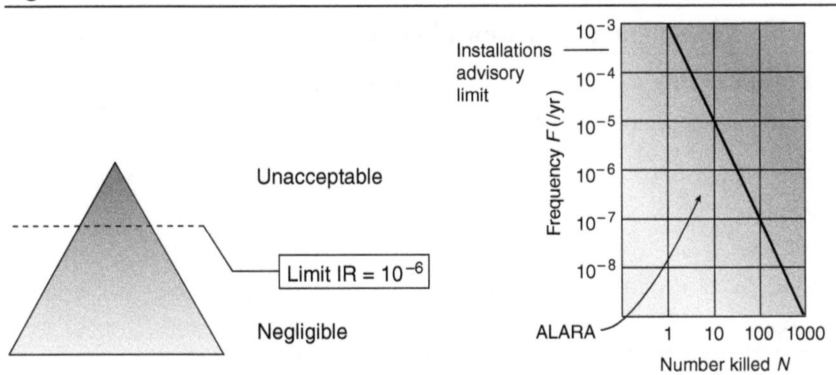

This risk value was reaffirmed when a decision had to be taken about the construction of the closure of the Oosterschelde estuary. To help preserve the ecosystems the design was changed from a closed solid dam to a movable barrier. This barrier was required to give the same protection as the dams.

Because of this background, the Dutch parliament had a history of debating safety in terms of probabilistic expectations, which came in handy when industrial risk had to be discussed. The accident in 1978 at an LPG road station prompted the chief inspector for the environment to take action. To limit the number of victims in the event of an explosion, limits were placed on the number of houses and offices that were allowed to be situated near to these stations (Table 6.3).

In the LPG study (TNO, 1983) it became apparent that individual risks from LPG installations fell sharply below the level of 10^{-6}. This was no surprise, in the sense that this level corresponds approximately to the value estimated at the time for the probability of catastrophic failure of the pressure tanks. This value corresponded conveniently with the value adopted by the Delta Commission, and thus it became the maximum acceptable addition to the risk of death for any individual resulting from such major accidents.

The value of one in a million per year also corresponds to about 1 per cent of the probability of being killed in a road accident in the mid-1980s. It was this value that was chosen as the maximum allowable individual risk for new situations.

For societal risk the anchor point was found in the 'interim viewpoint' relating to LPG points of sale. When combined with the value already chosen for individual risk, this led to the definition of the point corresponding to 10 people killed at a frequency of 1 in 100,000 per year. As societal risk is usually depicted as an FN curve having the frequency of exceeding *N* victims as a function of *N*, the limit had to be given the same form. Thus the slope of the limit line had to be determined.

A first attempt dates back to 1976: a document about environmental standards issued by the province of Groningen, since forgotten, set out limits on the acceptability (Figure 6.7) (Ball and Floyd, 1998). In the region where

Table 6.3 Building restrictions in relation to LPG fuel stations

Distance to tank	Building allowed	
or supply point (m)	Houses	Offices
0–25	None	None
25–50	Max. 2	Max. 10 people
50–100	Max. 8	Max. 30 people
110–150	Max. 15	Max. 60 people
>150	Free	Free

Source: Brief van de Hoofdinspecteur Milieuhygiene aan de gemeenten d.d. 24 mei 1977 en het Interimstandpunt LPG stations d.d. 28 april 1978, Vomil/|Vrom, the Netherlands

the consequences included people being killed, the slope was –2, as a representation of the aversion felt by people to large-scale accidents. It had a grey area of four orders of magnitude between the unacceptable and the negligible level. To a certain extent this approach was in turn based on an earlier document from the UK Atomic Energy Authority (UKAEA). In particular, Farmer (1967a and b) presented an acceptability curve for iodine-131 exposure whose (negative) slope gradually increased with increasing exposure.

It was decided to incorporate the apparent aversion to major disasters into the national limit by having the slope steeper than –1 there as well. An additional argument for this choice was that, at a slope of –1, the expectation value of the limit would be infinite. The only way to have a finite expectation value at a slope of –1 is to set an absolute maximum to the scale of an accident, which is not practically feasible, except for the fact that the maximum number of people who can be killed in any one accident is limited by the population of the earth (presently some 7 billion).

Several values circulated in the literature at the time, ranging from –1.2 to –2 (Farmer, 1967; Meleis and Erdman, 1972; Turkenburg, 1974; Wilson, 1975; Okrent, 1981; Rasbash, 1984; Smets, 1985; Hubert *et al.*, 1990). In the end it was decided to adopt a slope of –2 for the limit line in the Netherlands. In order to bound the decision space at the lower end of the risk spectrum, limits of negligibility were set for both individual risk and societal risk at 1 per cent of the value of the acceptability limit. The resulting complex of limit values was laid down in a policy document called 'Premises for Risk Management' (TK, 1988). Only a few other countries have published limits for external risks. Among these are the United Kingdom and Hong Kong (Ale, 1991, 1992, 1993).

Figure 6.7 Risk limits from the province of Groningen

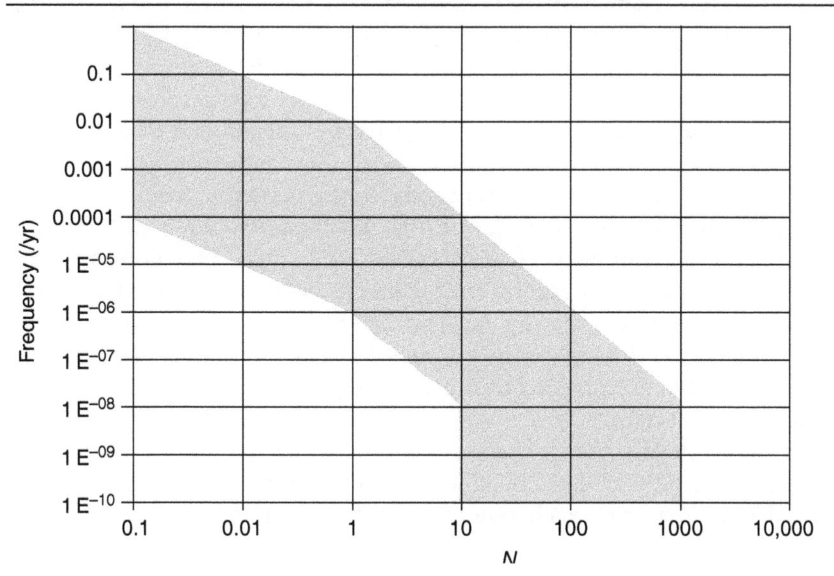

As we can see, the final choice is the result of practicality, convenience, political opportunity, common sense and the notion that protecting people from disasters is something worth doing. When risk managers are confronted with particular criteria, they therefore do well to ask themselves where these criteria came from and whether they are applicable in their situation.

Standstill

Political opportunity can also lead to the choice of 'standstill'. The current risk is accepted, but it cannot be allowed to grow worse. It is worth thinking about such a choice. It may look the easy way out of a difficult position, but it also makes the current situation the limit. In the case of Schiphol Airport, standstill of risk was promised to the population (MER, 1992; TK 2001; EK 2001) in spite of all the information indicating that the airport could grow only when the risks were allowed to grow also. After a plane actually crashed into the city of Amsterdam in 1992, it took more than 10 years before the government could admit that the promise could not be kept (Ale, 2006).

FRAMING

The answer to a question or the result of a decision-making process is sometimes heavily influenced by the way the problem is put – and sometimes by the possibilities of obtaining answers. We saw earlier that we obtain a different perspective when we look at the people who have lost their homes as a result of the mortgage crisis from the one we get than if we look at the bank that sold people mortgages to make fat profits. In this book we have looked at risk and looked at problems from a risk perspective. To look at a problem as a risk problem, you first have to accept that the problem is about risk. We have seen, too, that one way of defining the problem of nuclear power is as a problem concerning the sort of government you want and the amount of privacy you are prepared to give up to allow proper supervision of the fissile material.

You can define the problem of industrial risk as the problem of minimizing the number of statistical deaths. You can also start with the Old Testament commandment 'Thou shall not kill'. From that point of view, no level of risk is acceptable. Unfortunately, you then immediately run into the trouble of comparing evils, as many activities involve some probability of harm. Ethical positions such as that human life needs to be protected at all cost have led to decisions to go to war in which the loss of human life is certain.

In the light of these questions, decisions about minimizing or optimizing risk are relatively easy. But the decision maker has to be careful here too. The fact that it is much easier to calculate the number of people killed in an accidental explosion causes industrial risks to be expressed in numbers killed. The decision maker has to remain aware that there are also injured people to consider, and family, and society, and ... The most important message for the decision maker here is that what is put in front of him or her as a risk problem does not have to be one.

NUCLEAR WASTE

The amount of energy that can be obtained from the fission of nuclear material is enormous. However, most of the mass of the nuclear fuel remains in the form of waste.

When nuclear fuel is first placed in a commercial reactor, it consists of uranium oxide. After the fuel has been used for three or four years, it consists of about 96 percent uranium oxide and about 3 percent new elements – including iodine, strontium, carbon, xenon, caesium, silver and palladium. These are nuclear wastes.

Spent nuclear fuel is termed 'high-level waste', since it is considerably more radioactive than new fuel. A person can handle new fuel pellets of uranium oxide without danger. In contrast, spent fuel is dangerously radioactive, although much of the radioactivity dissipates quickly – some 98 percent of it within six months. Forty to 50 years after spent fuel is removed from the reactor, its radioactivity has decreased by a factor of 100. A very small percentage of nuclear wastes remains radioactive for thousands of years.

In addition to high-level waste, another category of nuclear waste associated with nuclear power is called 'low-level waste'. Low-level waste is generally anything that becomes contaminated with radioactive materials during its use. Such items include rags, papers, cleaning materials, protective clothing, tools and contaminated liquids. Uranium mine and mill tailings – the residue, sludges, and sands from uranium mining and milling operations – are also considered nuclear wastes because they contain very small concentrations of natural radioactive elements.

These wastes have to be stored until the radiation has diminished to a level where it poses no further danger to the environment. Typical storage times needed to get to these levels are of the order of 10,000 years. In Figure 6.8 a small impression is given of what has happened over the past 10,000 years.

The Great Pyramid in particular is a valuable clue. This pyramid was designed and built to keep the body of the Pharaoh for all eternity. But when it was opened the body had gone. And nobody really knows how the pyramid is put together; we can barely read the texts of the ancient Egyptians. And the building has been standing for only half the time span for which the waste needs to be kept and guarded.

Would the Egyptians have predicted our current society? What chance do we have to make a reliable prediction of what our world will look like

Figure 6.8 What has happened in the past 10,000 years

10,000 years ago	The first agriculture, planting wild grass with digging stick, begins in the Near East
8,000 years ago	The wheel invented in Sumeria
7,000 years ago	Copper smelted in Persia
6,000 years ago	Horses and cattle domesticated in the Indus Valley
5,000 years ago	The Great Pyramid of Cheops rises in Egypt, followed by the Great Sphinx
4,000 years ago	Stonehenge erected in England
3,000 years ago	The Parthenon erected in Athens; ships (supposedly) pass between the legs of the Colossus of Rhodes

10,000 years from now? Yet in order to keep the waste safe, we have to design something that will last that long. In a situation where we really do not have any facts and there is no way to know what will happen, we can only use our fantasy. In fact, we can only hope.

For these sorts of decisions neither a frequentist nor a Bayesian approach to probability and expectation will help, as both approaches need at least some sort of information for basing the estimates on.

DISCOURSE

For setting standards, for applying cost–benefit analyses and even more for making risk limits in law, there has to be reasonable certainty about the extent of the risk and about the costs of reducing it. There are also cases where the risk is virtually unknown. Problems such as global warming, or the potential detrimental effects of electromagnetic radiation, are difficult to quantify, if it can be done at all.

The most we can achieve may be an estimate of the potential detrimental consequences, and even that is difficult to make if the period of time over which these consequences have to be projected extends beyond a few decades.

The framing of problems and the decisions in uncertainty therefore need consultation and discussion with a wider audience than the decision makers and the consulting scientists alone. Discussion with 'all' stakeholders is called discourse. Such a discourse is intended to get as much relevant information and arguments on the table as possible before a decision is made. The advocates of problem solving via discourse expect that this would lead to wider acceptance of decisions. The opposers on the other hand are of the opinion that discourse is another attempt to railroad people into a decision.

Sometimes these discourses or negotiations lead to a decision. Sometimes this decision is taken for negotiated scientific truth. The decision sometimes is necessary, especially since no decision is often a decision as well. It is never necessary to forget the uncertainty or lack of knowledge.

Almost all the literature about this means of problem solving originates from rich, western, democratically governed countries. Decision makers – making decisions about risk, or any other subject – do well to realize that this is a very comfortable position of which the vast majority of humankind can only dream.

RISKY DECISIONS

We have seen that risks involve uncertainty about the immediate future, but not about the general behaviour of systems under consideration. Examples are the throwing of dice, driving a car and the transport of fuel. We have also seen that problems which are called risk problems are actually about uncertainty caused by a lack of understanding of the current or future behaviour of the system or, more fundamentally, by the impossibility of knowing the – distant – future. We have seen that the extent to which these problems can be addressed by systematic risk analysis and characterization is different and that that difference is important.

In any case, risk acceptability is a complex issue and in principle a political one. Only if the risk is confined to a financial institution or a financial endeavour can the risks be expressed in a single metric: money. In all other cases, things of a completely different nature have to be compared. People will value these things differently depending on their values, ethical position and beliefs. Politics is the only activity in which comparing apples and oranges is legitimate.

There are many attributes of a problem that influence people's decisions. The probability of an event is one of these, and an important one. Others are the extent of damage, the duration of the after-effects for the survivors, the temporal and geographical scale of the disaster, who will pay for the damages and when, and whether the public services will be ready to cope and capable of coping with the disaster and supporting the public. All these nuances are important for the stakeholders involved in the decision: public, experts, civil servants in policy branches of government, and politicians.

Information on quantifiable and also on non-quantifiable issues should all be input into the decision making. Experience in the field of chemical hazards has shown that decision making on quantified analysis using people killed as a proxy for the scale of potential disasters is a viable way of dealing with a difficult societal and political problem (Pikaar and Seaman, 1995).

Nevertheless, people will ask for more information, and in the long run, decisions based on numbers killed will be challenged on the grounds that additional information is not available or not explicitly taken into account (Hanemeier and den Hollander, 2003). Decisions involving human lives are always difficult and cannot be based solely on scientific or engineering facts, whatever the definition of facts in this context may be. In the case of an emergency it will be the population that will bear the consequences in the form of destruction of property and loss of life.

Compassion for people and willingness to listen to 'laypersons' are important qualities for authorities and public offices to possess. It would certainly help if the public authorities could metamorphose from nameless institutions into bodies peopled by human beings with feelings for their fellow citizens.

The bottom-line question, and therefore the sentence line of this book, is: 'If the accident happened tomorrow, would we then still think the risk acceptable?'

References

Agricola, G. (1556) *De metallica*, trans. H.C. Hoover and L.H. Hoover, The Mining Magazine London (1912).

Ale, B.J.M. (1991) Risk analysis and risk policy in the Netherlands and the EEC, *Journal of Loss Prevention in the Process Industries*, 4, 58–64.

Ale, B.J.M. (1992) The implementation of an external safety policy in the Netherlands, *Proceedings of the International Conference on Hazard Identification and Risk Analysis: Human Factors and Human Reliability in Process Safety*, Orlando Florida, 15–17 June, 173–184.

Ale, B.J.M. (1993) Dealing with risk of fixed installations in the Netherlands: Safety in the design and operation of low temperature systems, *Cryogenics*, 33 (8), 762–766.

Ale, B.J.M. (2001) The explosion of a fireworks storage facility and its causes, *Proceedings of the 2001 ASME IMECE*, New York, 11–16 November, IMECE2001/SERA-24014.

Ale, B.J.M. (2003) Dat overkomt ons niet, Oratie, TU Delft.

Ale, B.J.M. (2005) Tolerable or acceptable, a comparison of risk regulation in the United Kingdom and in the Netherlands, *Risk Analysis*, 25 (2), 231–104.

Ale, B.J.M. (2006) *The Occupational Risk Model*, Report of the Workgroup on ORM, TU-Delft, Risk Centre report no. RC20060731.

Ale, B.J.M. and Piers, M. (2000) The assessment and management of third party risk around a major airport, *Journal of Hazardous Materials*, 71 (1–3), 1–16.

Ale, B.J.M. and Uitdehaag, P.A.M. (1999) *Guidelines for Quantitative Risk Analysis* (CPR18), RIVM, SDU, The Hague.

Ale, B.J.M. and Whitehouse, R. (1984) A computer based system for risk analysis of process plants. In *Heavy Gas and Risk Assessment III*, S. Hartwig (ed.), D. Reidel, Dordrecht, the Netherlands.

Ale, B.J.M., Bellamy, L.J., Cooke, R.M., Goossens, H.J. Hale, A.R., Roelen, A.L.C. and Smith, E. (2006) Towards a causal model for air transport safety – an ongoing research project, *Safety Science*, 44 (8), 657–673.

Ale, B.J.M., Bellamy, L.J., van der Boom, R., Cooper, J., Cooke, R.M., Goossens, L.H.J., Hale, A.R., Kurowicka, D., Morales, O., Roelen, A.L.C and Spouge, J. (2007) Further development of a causal model for air transport safety (CATS): building the mathematical heart, in T. Aven and J.E. Vinnem (eds), *Risk, Reliability and Societal Safety*, Taylor & Francis, London.

Ale, B.J.M., Bellamy, L.J., van der Boom, R., Cooper, J., Cooke, R.M., Goossens, L.H.J., Hale, A.R., Kurowicka, D., Morales, O., Roelen, A.L.C. and J. Spouge, (2008) *Development of a Causal Model for Air Transport Safety (CATS): Final Report*, Ministry of Transport and Water Management, The Hague.

Amalberti, R. (2001) The paradoxes of almost totally safe transport systems, *Safety Science*, 37, 109–126.

Arnaud, A. (1662) *La Logique, ou l'art de penser.*

Arshinov, V. and Fuchs, C. (eds) (2003) *Causality, Emergence, Self-Organisation*, http://www.self-organization.org/results/book/EmergenceCausalitySelf-Organisation.pdf.

Asselt, M.B.A. van (2000) *Perspectives on Uncertainty and Risk: the PRIMA Approach to Decision Support*, Kluwer, Dordrecht, the Netherlands.

Bahwar, I. and Trehan, M. (1984) Bhopal: city of death, *India Today*, 31 December, p. 4–25.

Ball, D.J. and Floyd, P. J. (1998) *Societal Risks, Final Report to HSE*, HSE Books, London.

Bayes, T. (1763) An essay towards solving a problem in the doctrine of chances, *Philosophical Transactions*, Essay LII, 370–418.

Beck, U. (1986) *Risikogesellschaft. Auf dem Weg in eine andere Moderne*, Suhrkamp, Frankfurt am Main. English translation: Risk Society: Towards a New Modernity, Sage, London (1992).

Bedford, T. and Cooke, R. (2001) *Probabilistic Risk Analysis: Foundations and Methods*, Cambridge University Press, Cambridge.

Bellamy, L.J. (1986) *Review of Evacuation Data.* Technica Report C768. Prepared for the Ministry of Housing, Spatial Planning and Environment (VROM), the Netherlands.

Bellamy, L.J. (1987) *Evacuation of the Public in the Vicinity of a Nuclear Installation.* Technica Report C1034/204 prepared for Her Majesty's Nuclear Installations Inspectorate, UK.

Bellamy, L.J. (1989a) *Investigation of the Usage of Sirens as an Offsite Public Warning Medium.* Technica Report C1719, prepared for Her Majesty's Nuclear Installations Inspectorate, UK.

Bellamy, L.J. (1989b) *An Evaluation of Sheltering as a Mitigating Response to Radioactive Releases.* Technica Report C1696, prepared for Her Majesty's Nuclear Installations Inspectorate, UK.

Bellamy, L.J. (1989c) Informative fire warning systems: a study of their effectiveness, *Fire Surveyor*, 18 (4), 17–23.

Bellamy, L.J. (1993) Where human factors science can support offshore search and rescue, paper presented to the 12th Leith International Conference and Exhibition on Offshore Search and Rescue, 01–04 November.

Bellamy, L.J. (2000) Risico-analyses t.b.v. MAVIT (Risk analyses in relation to Societal Acceptance of Risk in Tunnels), SAVE Report 000093 – G77, Apeldoorn, The Netherlands, prepared for Ministry of Home Affairs (Ministerie BZK), Directorate Fire Brigades and Disaster Abatement.

Bellamy, L.J. and Geyer, T.A.W (1990) *Experimental Programme to Investigate Informative Fire Warning Characteristics for Motivating Fast Evacuation*, Building Research Establishment Report BR 172.

Bellamy, L.J. and Harrison, P.I. (1988) An evacuation model for major accidents, *Proceedings of Disasters and Emergencies: The Need for Planning*, IBC Technical Services, London.

Bellamy, L.J., Bacon, L.G., Geyer, T.A.W and Harrison, P.I. (1987) *Human Response Modelling for Toxic Releases: Model Theory prepared for the Ministry of Housing, Spatial Planning and Environment (VROM)*, the Netherlands, Technica Report C944.

Bellamy, L.J., Geyer, T.A.W., Max-Lino, R., Harrison, P.I., Bahrami, Z. and Modha, B. (1988) An Evaluation of the Effectiveness of the Components of Informative Fire Warning Systems, In J.D. Simes (ed.) *International Conference on Safety in the Built Environment*, E. & F. Spon, London.

Bellamy, L.J. and Brabazon, P.G. (1993) Problems of Understanding Human Behaviour in Emergencies, Invited paper presented at Response to Incidents Offshore, 8/9 June 1993, Aberdeen, organised by IBC Technical Services Ltd.

Bellamy, L.J., Oh, J.I.H., Hale, A.R., Papazoglou, I.A., Ale, B.J.M., Morris, M., Aneziris, O., Post, J.G., Walker, H., Brouwer, W.G.J. and Muyselaar, A.J. (1993) IRISK: ' Development of an Integrated Technical and Management Risk Control and Monitoring Methodology for Managing On-Site and Off-Site Risks, EC Contract Report ENV4-CT96-0243 (D612-D).

British Railways Board (BRB) (1994) *Automatic Train Protection*. Report from the BRB to the Secretary of State for Transport, BRB.

Button, K. (1993) *Overview of Internalising the Social Costs of Transport*, OECD ECMT report.

CBS (2006) STATLINE, CBS, www.cbs.nl.

Cerf, C. and Navasky, V. (1984) *The Experts Speak: The Definitive Compendium of Authoritative Misinformation*, Pantheon, New York.

Chalmers, D.J. (1996) *The Conscious Mind: In Search of a Fundamental Theory*, Oxford University Press, Oxford.

Chaucer (1387–1400) *Canterbury Tales*, Chancellor, London (1985).

Chemical Industries Association (CIA) (1977) *A Guide to Hazard and Operability Studies*, Alembic House, London.

Chilton, S., Covey, J., Hopkins, L., Jones-Lee, M., Loomes, G., Pidgeon, N. and Spencer, A. (1998) New research on the valuation of preventing fatal road accident casualties, in Department for the Environment, Transport and the Regions (ed.) *Road Accidents Great Britain*, The Stationery Office, London.

Cooke, R.M. and Goossens, L.J.H. (2000) *Procedures Guide for Structured Expert Judgement*, European Commission, EUR 18820 EN.

Cox, R.A. and Bellamy, L.J. (1992) The use of QRA and human factors science in emergency preparedness and crisis management, paper presented at Risk Analysis and Crisis Management, the Interface, 22–23 September 1992, London. Organized by BPP Technical Services/Cremer & Warner.

CPR (1990) *Methoden voor het berekenen van Schade* (CPR14), Directoraat Generaal van de Arbeid, SDU, The Hague.

Cremer & Warner (1981) *Risk Analysis of Six Potentially Hazardous Industrial Objects in the Rijnmond Area*, D. Reidel, Dordrecht, the Netherlands.

Cullen, Lord W. (2001) *The Ladbroke Grove Rail Inquiry*, The Stationery Office, London.

Debray, B., Delvosalle, C., Fiévez, C., Pipart, A., Londiche, H. and Hubert, E. (2004) Defining safety functions and safety barriers from fault and event trees analysis of major industrial hazards, C. Spitzer, U. Smocker and V.N. Dang (eds) *Proceedings of PSAM7*, Springer-Verlag, London.

Dekker, S. (2006) *A Field Guide to Understanding Human Error*, Ashgate, Aldershot, UK.

Delvosalle, C., Fievez, C. and Pipart, A. (2004) Report presenting the final version of the methodology for the identification of reference accident scenarios. ARAMIS Project – fifth Framework Program of the European Community, Mons, Belgium, July.

Delta commissie (1960) *Rapport van de Deltacommissie*, Delta wet 1957.
Department of Transport (DoT) (1995) Mawhinney endorses HSC view on future of Automatic Train Protection, DoT Press Notice 98, 30 March.
Dhillon, B.S. (1992) Failure mode and effects analysis: bibliography, *Microelectronics and Reliability*, 32, 719–731.
Dijksterhuis A. (2007) *Het Slimme Onbewuste*, Bert Bakker, Amsterdam, ISBN-13: 9789035129689.
DNV (2008) *Fault Tree Modelling for the Causal Model of Air Transport Safety Final Report for Ministerie van Verkeer en Waterstaat DNV Project* No. C21004587/3 Revision 0 20 June 2008, DNV, London.
Drieu, C. - Project leader (1995) *Can the processes of mental representation and decision-making in a major emergency or crisis situation be modified? From a symbolic crisis to a crisis in progress.* Final report for the EU. Contract number CT 94 0090.
Einstein, A. and Infeld, L. (undated) *The Evolution of Physics*, The Cambridge Library of Modern Science, Cambridge University Press, London.
Eisenberg, Norman A., Lynch, Cornelius J., and Breeding, Roger J. (1975) *Vulnerability Model. A Simulation System for Assessing Damage Resulting from Marine Spills*, Enviro control inc rockville md.
EK (2001) Records of the First Chamber of Parliament of the Netherlands, year 2001–2003, 27603 no. 88k.
EU (1982) European Union Directives 82/501/EEG (Pb EG 1982, L 230) and 87/216/EEG (Pb EG 1987, L85) ("Seveso directive").
Evans, A.W (1996) The economics of automatic train protection in Britain, *Transport Policy*, 3 (3), 105–110.
Evans, A.W. (2005) Railway risks, safety values and safety costs, *Proceedings of the Institution of Civil Engineers Transport*, 158 (1) 3–9.
Farmer, F.R. (1967a) Reactor safety and siting: a proposed risk criterion, *Nuclear Safety*, 8, 539.
Farmer, F.R. (1967b) Siting criteria: a new approach, *Atom*, 128, 152–179.
Gezondheidsraad (1993) *Risico is meer dan een getal*, Gezondheidsraad, The Hague.
Groeneweg, J. (1998) *Controlling the Controllable*, DSWO Press, Leiden.
Haddon, W. Jr (1973) Energy damage and the 10 countermeasure strategies, *Human Factors*, 15 (4), 355–366.
Hanemeier P. and den Hollander, G. (2003) Nuchter omgaan met risico's (Coping Rationally with Risks), Milieu- en Natuurplanbureau (MNP), RIVM report 251701047/2003, RIVM, Bilthoven, the Netherlands.
Harrison, P.I. and Bellamy, L.J. (1988) Modelling the evacuation of the public in the event of toxic releases: a decision support tool and aid for emergency planning, in B.A. Sayers (ed.) *Human Factors and Decision Making: Their Influence on Safety and Reliability*, Elsevier, London.
Heinlein, R.A. (1973) *Time Enough for Love*, Putnam, New York.
Heinrich H. W. (1959). *Industrical accident prevention: a scientific approach*, 4th ed., McGraw-Hill.
Hidden, A. (1989) *Investigation into the Clapham Junction Railway Accident*, HMSO, London.
HSE (Health and Safety Executive) (1978) *Canvey: An Investigation of Potential Hazards from Operations in the Canvey Island/Thurrock Area*, HMSO, London.

HSE (1981) *Canvey: Second Report: A Review of the Potential Hazards from Operations in the Canvey Island/Thurrock Area Three Years after Publication of the Canvey Report*, HMSO, London.
HSE (2000) *The Train Collision at Ladbroke Grove 5 October 1999: A Report of the HSE Investigation*, The Stationery Office, London.
HSE (2001) *Reducing Risk, Protecting People*, The Stationery Office, London.
HSE (2002) *Policy, Risk and Science: Securing and Using Scientific Advice*, HSE Contract Research Report 595/2000, HSE Books, London.
House of Commons, Science and Technology Committee (2006) *Scientific Advice, Risk and Evidence Based Policy Making*, The Stationary Office, London.
http(1)://en.wikipedia.org/wiki/World_energy_resources_and_consumption per 10/08/2008.
Hollnagel, E. (2006) *Barriers and Accident Prevention*, Ashgate, Aldershot, UK.
Houtenbos, M. (2008) Expecting the unexpected, PhD thesis, Technical University Delft, SWOV, Leidschendam, the Netherlands.
Hubert, Ph., Barni, M.H. and Moatti, J.P. (1990) Elicitation of criteria for management of major hazards, paper presented at Second SRA Conference, Laxenburg, Austria, 02–03 April.
Hudson, P. (2001) Safety culture: the ultimate goal, *Flight Safety Australia*, 5 (5), September–October, 29–31.
ICI (1979) Adapted from ICI Plc Hazan Course Notes.
International Civil Aviation Organization (ICAO) (2000) *Accident/Incident Reporting Manual* (ADREP) ICAO, Montreal.
Jongejan, R.B., Ale, B.J.M. and Vrijling, J.K. (2006) FN-criteria for risk regulation and probabilistic design, paper presented at International Conference on Probabilistic Safety Assessment and Management, 13–19 May, New Orleans.
Jongerius, R.T. (1993) *Spoorwegongevallen in Nederland*, Schuyt, Haarlem, the Netherlands.
Kant, I. (1783) *Kritik der reinen Vernunft*, Riga.
Kaplan, S. and Garrick, B.J. (1981) On the quantitative definition of risk, *Risk Analysis Journal*, 1 (1), 11–27.
Kjellen, U. (1983) Analysis and development of corporate practices for accident control, thesis, Royal Institute of Technology, Report no. Trita AVE-0001, Stockholm.
Knight, F.H. (1921) *Risk, Uncertainty and Profit*, reissued by Cosimo Classics, New York (2005).
Kuhlman, A. (1981) *Einfürung in die Sicherheitswissenchaft*, Friedr. Vieweg, Cologne.
Leeuwen, C.J. van and Hermens, J.L.M. (eds) (1995) *Risk Assessment of Chemicals: An Introduction*, Kluwer, Dordrecht, the Netherlands.
Lin, P.H., Hale, A.R., Gulijk, C. van, Ale, B.J.M., Roelen, A.L.C. and Bellamy, L.J. (2008) Testing a safety management system in aviation, *Proceedings of the Ninth International Probabilistic Safety Assessment and Management Conference*, Hong Kong, 18–23 May.
MacDonald, G.L. (1972) *The involvement of tractor design in accidents, Research report* 3/72, Department of Mechanical Engineering, University of Queensland, St Lucia, 1972.
McCarthy, M.P. and Flynn, T.P. (2004) *Risk from the CEO and Board Perspective*, McGraw-Hill, New York.

Mann, M.E., Bradley, R.S. and Hughes, M.K. (1999) Northern Hemisphere temperatures during the past millennium: inferences, uncertainties, and limitations, *Geophysical Research Letters*, 26, 759–762.

Maslow, A.H. (1943) A theory of human motivation, *Psychological Review*, 50, 370–396.

Maslow, A.H. (1954) *Motivation and Personality*, Harper & Row, New York.

Meleis, M. and Erdman, R.C. (1972) *The development of reactor siting criteria based upon risk probability*, Nuclear Safety, 13, 22.

MER (1992) *Commissie MER, richtlijnen voor de MER Schiphol*, February.

Morall, J.F. III (1986) A review of the record, *Regulation*, 10 (2), 25–34.

Morall, J.F. III (1992) Controlling regulatory costs: the use of regulatory budgeting, Regulatory Management and Reform Series no. 2, OECD/GD(92)176.

Mosleh, A., Fleming, K.N., Parry, G.W., Paula, H.M., Rasmuson, D.M. and Worledge, D.H. (1988) *Procedures for Treating Common Cause Failures in Safety and Reliability Studies*, NUREG/CR 4780, US Regulatory Commission, 2 vols (1988 and 1989).

Nash, J.R (1976) *Darkest Hours*, Nelson Hall, Chicago.

National Transportation Safety Board (NTSB) (2000) Aircraft accident report *PB2000-910403 NTSB/AAR-00/03 DCA96MA070, In-flight Breakup over the Atlantic Ocean Trans World Airlines Flight 800 Boeing 747–131, N93119, Near East Moriches, New York, July 17, 1996, 6788G*, adopted 23 August 2000.

Nickell, R., Jones, J.W. and Balkey, K.R. (2004) RAMCAP: Risk Analysis and Management for Critical Asset Protection, overview of methodology, paper presented at AMSE International Mechanical Engineering Congress and RD7D Expo, Annaheim, CA, 13–19 November.

NN (1933) Wat ooggetuigen over de ramp te Leiden in 1807 vertellen, in *Jaarboekje voor Geschiedenis en Oudheidkunde van Leiden en Rijnland* 25, pp. 1–13 (met verbetering in het jaarboekje van 1944, p. 215).

NN (1949) *Edwards vs. The National Coal Board* 1 A11 ER 743.

NN (1977) Brief van de Hoofdinspecteur Milieuhygiene aan de gemeenten d.d. 24 mei 1977 en het Interimstandpunt LPG stations d.d.28 april 1978, Vomil/VROM, the Netherlands.

NN (1995) ExternE, *Externalities of Energy, vol. 2, Methodology*, European Commission, EUR 16521 EN.

OBL (2007) *Gebroken Hart*, Gemeente Amsterdam.

OECD (2003) *Emerging Risks in the 21st Century*, An OECD International Futures Project, September.

Oh, J.H., Brouwer, W.G.J., Bellamy, L.J., Hale, H.R., Ale, B.J.M. and Papazoglou, J.A. (1998) The Irisk Project: development of an integrated technical and management risk control and monitoring methodology for managing and quantifying on-site and off-site risk, in A. Mosleh and R.A. Bari (eds) *Probabilistic Safety Analysis and Management*, vol. 4, (PSAM4), Springer, New York.

Okrent, J. (1981) Industrial risk, Proceedings of the Royal Society, 372, 133–149.

Oosting, M. (2001) *Eindrapport van de commissie onderzoek vuurwerkramp*, Sdu, The Hague.

Otway, H.J. (1973) Risk estimation and evaluation, in *Proceedings of the IIASA Planning Conference on Energy Systems*, IIASA-PC-3, International Institute of Allied Systems Analysis, Laxenburg, Austria.

Otway, H.J. (1975) Risk assessment and social choices, IIASA Research Memorandum, International Institute of Allied Systems Analysis, Laxenburg, Austria.

OVV (2006) Brand Cellencomplex Oost, Project no. M2005SCH1026-1, The Hague, 21 September.

Papazoglou, I.A. and Ale, B.M. (2007) A logical model for quantification of occupational risk, Reliability Engineering and System Safety, 92 (6), 785–803.

Papazoglou, I.A., Bellamy, L.J., Hale, A.R., Aneziris, O.N., Ale, B.J.M., Post, J.G. and Oh, J.I.H. (2003) I-Risk: development of an integrated technical and management risk methodology for chemical installations, *Journal of Loss Prevention in the Process Industries*, 16, 575–591.

Papazoglou, I.A., Aneziris, O., Post, J., Baksteen, H., Ale, B.J.M., Oh, J.I.H, Bellamy, L.J., Mud, M.L., Hale, A., Goossens, L. and Bloemhoff, A. (2006) Logical models for quantification of occupational risk: falling from mobile ladders, paper presented at the International Conference on Probabilistic Safety Assessment and Management, 13–19 May, New Orleans.

Pikaar, M.J. and Seaman, M.A. (1995) *A Review of Risk Control*, Report no. SVS 1994/27A, Ministry of Housing, Physical Planning and Environment, the Netherlands.

Ramazzini, B. (1700) *De morbis artificum deatriba.*

Rasbash, D.J. (1984) Criteria for acceptability for use with quantitative approaches to fire safety, *Fire Safety Journal*, 8, 141–158.

Rasmussen, J. (1997) Risk management in a dynamic society: a modeling problem, *Safety Science*, 27 (2/3), 183–213.

Reason, J. (1990) *Human Error*, Cambridge University Press, Cambridge.

Reason, J. (1997) *Managing Risks of Organizational Accidents*, Ashgate, Aldershot, UK.

Reniers, G.L.L., Dullaet, W., Ale, B.J.M., Vverschuren, F. and Soudan, K. (2007) Engineering an instrument to evaluate safety critical manning arrangements in chemical industrial areas, *Journal of Business Chemistry*, 4 (2), 60–75.

Reniers, G.L.L., Pauwels, N., Audenaert, A., Ale, B.J.M. and Soudan, K. (2008) A multiple shutdown method for managing evacuation in case of fire accidents in chemical clusters, *Journal of Hazardous Materials*, 152 (2), 750–756.

RIVM (2006) Nationaal Kompas Volksgezondheid, at www.rivm.nl.

Rosen, G. (1976) *A History of Public Health*, MD Publications, New York.

Russell, B. (1946) *A History of Western Philosophy*, George Allen & Unwin, London.

Schupp, B.A., Smith, S.P., Wright, P. and Goossens, L.H.J. (2004) Integrating human factors in the design of safety critical systems: a barrier based approach, *Human Error, Safety and Systems Development*, 152, 285–300.

Sjoberg, L. (2000) Factors in risk perception, *Risk Analysis*, 20 (1) 1–11.

Slovic, P. (1999) Emotion, sex, politics and science: surveying the risk assessment battlefield, *Risk Analysis*, 19 (4), 689–701.

Slovic, P., Fischoff, B., Lichtenstein, S., Read, S. and Combs, B. (1978) How safe is safe enough? A psychometric study of attitudes towards technological risks and benefits, *Policy Sciences*, 8: 127–152.

Smets, H. (1985) Compensation for exceptional environmental damage caused by industrial activities, paper presented at Conference on Transportation, Storage and Disposal of Hazardous Materials, International Institute of Applied Systems Analysis, Laxenburg, Austria.

Spouge, John (2008) *Fault Tree Modelling for the Causal Model of Air Transport Safety Final Report*, DNV London, DNV Project No. C21004587/3, June 2008.

Stewart, I. (2002) *Does God Play Dice? The Mathematics of Chaos*, 2nd edn. Malden, MA: Blackwell.

Svedung, I. and Rasmussen, J. (2002) *Safety Science*, 40, 397–417.

Swain, A.D. and Guttman, H.E. (1983) *Handbook of Human Reliability Analysis with Emphasis on Nuclear Power Plants Applications*, NUREG/CR-1278 Sandia National Laboratories.

Tengs, T.O., Adams, M.E., Pliskin, J.S., Safran, D.G., Siegel, J.E., Weinstein, M. and Graham J.D. (1995) Five hundred life saving interventions and their cost effectiveness, *Risk Analysis*, 15, 369–390.

Thucydides (431 BC) *The Peloponnesian War, Pericles' Funeral Oration*, trans. R. Warner, Penguin Books, Harmondsworth, UK (revised edn (1972)).

TK (1988) *Omgaan met Risico's, Tweede Kamer, vergaderjaar 1988–1989*, 21137, no. 5. An English translation with the same number is titled *Premises for Risk Management.*

TK (2001) Records of the Second Chamber of Parliament of the Netherlands, year 2001–2003, 27603 no. 52.

TNO (1983) *A Comparative Analysis of the Risks Inherent in the Storage, Transshipment, Transport and Use of LPG and Motorspirit*, VROM, the Netherlands.

TRC (2003) *Buckling up; Technologies to increase seat belt use.* Special Report 278. Transportation Research Board TRB, National Research Council NRC. Washington DC.

Turkenburg, W.C. (1974) Reactorveiligheid en risico-analyse, *De Ingenieur*, 86 (10) 189–192.

Unwin, S.D. (2003) *The Probability of God,* Three Rivers Press, New York.

UNEP (2001) United Nations Environmental Programme, Intergovernmental Panel on Climate Change (IPCC) *Third Assessment Report: Climate Change 2001*, IPCC, Geneva.

Vesely, W.E. (1981) *The Fault Tree Hanboek*, F.F. Goldberg, N.H. Roberts, University of Washington, D.F. Haasl, NUREG-0492 U.S. Nuclear Regulatory Commission, Washington, DC 20555.

Viscusi, W.K. and Aldy, J. (2003) The value of a statistical life: a critical review of market estimates throughout the world, *Journal of Risk and Uncertainty*, 27, 5–76.

Visser, J.P. (1998) "Developments in HSE management in oil and gas exploration and production", in A.R. Hale and M. Baram (eds) *Safety Management: The Challenge of Change*, Pergamon, Oxford.

Vlek, C. (1996) A multi-stage, multi-level and multi-attribute perspective on risk assessment decision making and risk control, *Risk Decision Policy*, 1 9–31.

Willett, A.H. (1901) *The Economic theory of Risk and Insurance*, University of Pennsylvania Press, Philadelphia, reissued by University Press of the Pacific, Honolulu.

Wilson, R. (1975) The cost of safety, *New Scientist*, 68, 274–275.

Index